DISCARDED

GOOD SCIENCE

INSIDE TECHNOLOGY

EDITED BY WIEBE E. BIJKER, W. BERNARD CARLSON, AND TREVOR PINCH

Janet Abbate, *Inventing the Internet*

Atsushi Akera,*Calculating a Natural World: Scientists, Engineers and Computers during the Rise of U.S. Cold War Research*

Morana Alač, *Handling Digital Brains: A Laboratory Study of Multimodal Semiotic Interaction in the Age of Computers*

Stathis Arapostathis and Graeme Gooday, *Patently Contestable: Electrical Technologies and Inventor Identities on Trial in Britain*

Charles Bazerman, *The Languages of Edison's Light*

Marc Berg, *Rationalizing Medical Work: Decision-Support Techniques and Medical Practices*

Wiebe E. Bijker, *Of Bicycles, Bakelites, and Bulbs: Toward a Theory of Sociotechnical Change*

Wiebe E. Bijker and John Law, editors, *Shaping Technology/Building Society: Studies in Sociotechnical Change*

Wiebe E. Bijker, Roland Bal, and Ruud Hendricks, *The Paradox of Scientific Authority: The Role of Scientific Advice in Democracies*

Karin Bijsterveld, *Mechanical Sound: Technology, Culture, and Public Problems of Noise in the Twentieth Century*

Stuart S. Blume, *Insight and Industry: On the Dynamics of Technological Change in Medicine*

Pablo J. Boczkowski, *Digitizing the News: Innovation in Online Newspapers*

Geoffrey C. Bowker, *Memory Practices in the Sciences*

Geoffrey C. Bowker, *Science on the Run: Information Management and Industrial Geophysics at Schlumberger, 1920–1940*

Geoffrey C. Bowker and Susan Leigh Star, *Sorting Things Out: Classification and Its Consequences*

Louis L. Bucciarelli, *Designing Engineers*

Michel Callon, Pierre Lascoumes, and Yannick Barthe, *Acting in an Uncertain World: An Essay on Technical Democracy*

H. M. Collins, *Artificial Experts: Social Knowledge and Intelligent Machines*

Park Doing, *Velvet Revolution at the Synchrotron: Biology, Physics, and Change in Science*

Paul N. Edwards, *The Closed World: Computers and the Politics of Discourse in Cold War America*

Andrew Feenberg, *Between Reason and Experience: Essays in Technology and Modernity*

Tarleton Gillespie, Pablo J. Boczkowski, and Kirsten A. Foot, editors, *Media Technologies: Paths Forward for Social Research*

Michael E. Gorman, editor, *Trading Zones and Interactional Expertise: Creating New Kinds of Collaboration*

Herbert Gottweis, *Governing Molecules: The Discursive Politics of Genetic Engineering in Europe and the United States*

Joshua M. Greenberg, *From Betamax to Blockbuster: Video Stores and the Invention of Movies on Video*

Kristen Haring, *Ham Radio's Technical Culture*

Gabrielle Hecht, *Entangled Geographies: Empire and Technopolitics in the Global Cold War*

Gabrielle Hecht, *The Radiance of France: Nuclear Power and National Identity after World War II*

Kathryn Henderson, *On Line and On Paper: Visual Representations, Visual Culture, and Computer Graphics in Design Engineering*

Christopher R. Henke, *Cultivating Science, Harvesting Power: Science and Industrial Agriculture in California*

Christine Hine, *Systematics as Cyberscience: Computers, Change, and Continuity in Science*

Anique Hommels, *Unbuilding Cities: Obduracy in Urban Sociotechnical Change*

Deborah G. Johnson and Jameson W. Wetmore, editors, *Technology and Society: Building Our Sociotechnical Future*

David Kaiser, editor, *Pedagogy and the Practice of Science: Historical and Contemporary Perspectives*

Peter Keating and Alberto Cambrosio, *Biomedical Platforms: Reproducing the Normal and the Pathological in Late-Twentieth-Century Medicine*

Eda Kranakis, *Constructing a Bridge: An Exploration of Engineering Culture, Design, and Research in Nineteenth-Century France and America*

Jens Lachmund, *Greening Berlin: The Co-Production of Science, Politics, and Urban Nature*

Christophe Lécuyer, *Making Silicon Valley: Innovation and the Growth of High Tech, 1930–1970*

Pamela E. Mack, *Viewing the Earth: The Social Construction of the Landsat Satellite System*

Donald MacKenzie, *Inventing Accuracy: A Historical Sociology of Nuclear Missile Guidance*

Donald MacKenzie, *Knowing Machines: Essays on Technical Change*

Donald MacKenzie, *Mechanizing Proof: Computing, Risk, and Trust*

Donald MacKenzie, *An Engine, Not a Camera: How Financial Models Shape Markets*

Cyrus C. M. Mody, *Instrumental Community: Probe Microscopy and the Path to Nanotechnology*

Maggie Mort, *Building the Trident Network: A Study of the Enrollment of People, Knowledge, and Machines*

Peter D. Norton, *Fighting Traffic: The Dawn of the Motor Age in the American City*

GOOD SCIENCE
THE ETHICAL CHOREOGRAPHY OF STEM CELL
RESEARCH

CHARIS THOMPSON

THE MIT PRESS
CAMBRIDGE, MASSACHUSETTS
LONDON, ENGLAND

MIT Press books may be purchased at special quantity discounts for business or sales promotional use. For information, email special_sales@mitpress.mit.edu.

Set in Bembo by the MIT Press. Printed and bound in the United States of America.

Library of Congress Cataloging-in-Publication Data

Thompson, Charis.
Good science : the ethical choreography of stem cell research / Charis Thompson, Inside Technology.
  pages   cm
Includes bibliographical references and index.
ISBN 978-0-262-02699-4 (hardcover : alk. paper) 1. Stem cells—Research—Moral and ethical aspects—United States. 2. Stem cells—Research—Government policy—California. 3. Stem cells—Research—United States—Finance. 4. Federal aid to medical research—United States. I. Title.
QH588.S83.T48 2013
[R853.H8]
616.02'774—dc23
2013024261

10   9   8   7   6   5   4   3   2   1

CONTENTS

ACKNOWLEDGMENTS

I thank family, friends, colleagues, and students for their inspiring intellectual work and for their commitment to the issues of the day. I am especially fortunate to have worked with the faculty, staff, and students of the units that together make up the Gender and Women's Studies Department, the Center for Science, Technology, and Medicine in Society, and the Center for Race and Gender at UC Berkeley. I am grateful to the Deans of the Social Sciences and Humanities, the Committee on Research at UC Berkeley, and the Mellon Foundation for supporting the research upon which this book is based. I also much appreciate the support from the California Institute for Regenerative Medicine and from UC Berkeley's former Chancellor, Robert Birgeneau, and Executive Vice Chancellor, George Breslauer, for the Stem Cells and Society program.

I would also like to acknowledge the UC Berkeley Stem Cell Center, the Center for Genetics and Society, Generations Ahead, the Greenlining Institute, Sins Invalid, the Children's Hospital Oakland Research Institute, and the UC Berkeley ESCRO committee, with certain of whose members I had the pleasure of working and talking over several years. I thank the patients, donors, doctors, scientists, patient advocates, ethicists, historians, sociologists, lawyers, theologians, anthropologists, philosophers, business women and men, members of the press, students, and other members of the public with whom I talked and worked while preparing the manuscript.

For conversations germane to the book I would like, in particular, to thank Vincanne Adams, Adrienne Asch, Patricia Berne, Bob Birgeneau, Peter Braude, Philip Campbell, Alta Charo, Nancy Chen, Anne Cheng, Richard Chetkowski, Cho Hae-Joang, Adele Clarke, Lawrence Cohen, Irina Conboy, Marcy Darnovsky, Joe Dumit, Troy Duster, Jill Einstein,

Steven Epstein, Anne Fausto-Sterling, Marie Fox, Sarah Franklin, Peter Galison, Hank Greely, Zach Hall, Donna Haraway, Cori Hayden, Klaus Hoeyer, Lisa Ikemoto, Marie-Andrée Jacob, Sheila Jasanoff, Donna Jones, Kim Eunshil, Kim Sang-Hyun, Bob Klein, Geoff Lomax, Bert Lubin, Jennifer McCormick, Lily Mirels, Robin Mitchell, Michelle Murphy, Alondra Nelson, Osagie Obasogie, Aihwa Ong, Michael Pollan, Bob Price, Jack Price, Radhika Rao, Paul Rabinow, Jenny Reardon, Dorothy Roberts, Liz Roberts, Nikolas Rose, Randy Schekman, Nancy Scheper-Hughes, Silke Schicktanz, Brian Smith, Marilyn Strathern, Ken Taymor, Robert Tjian, Steven Wainwright, Clare Williams, David Winickoff, Wu Chia-Ling, and Yoon Jeong-Ro.

While doing the research for the book, I worked with the following students and postdoctoral fellows on related topics: Laurel Barchas, Katherine Darling, Krishanu Saha, Joe Tayag, and Aba Yamoah (all then undergraduates), Ruha Benjamin, Angie Boyce, Katherine Chandler, Alenda Chang, Jessica Davies, Jason Delbourne, Kate Drabinski, Carrie Friese, Chris Ganchoff, Orit Halpern, Natasha Hammond-Browning, Katie Hasson, Ben Hickler, Kerry Holden, Anna Jabloner, Zakiyyah Jackson, Elizabeth Jenner, Sujatha Jesudason, Jason Kim, Jennifer Liu, Jennifer Lorvick, Jennifer Lum, Lowry Martin, Kate Mason, Neide Mayumi Osada, Betty Padilla, Eric Plemons, Christine Quinan, Kelly Rafferty, Margaret Rhee, Chris Roebuck, Thurka Sangaramoorthy, Veronica Sanz, Krista Sigurdson, Anthony Stavrianakis, Rasheed Tazudeen, Annika Thiem, IK Udekwu, Jami Xu, and Yoon Hyaesin (all then graduate students), and Jake Kosek, Kim TallBear, Elly Teman, and Martin Weiss (all then postdoctoral fellows). As ever, the advising process enriched research.

I received extremely helpful, generous, and generative comments on an early draft from anonymous reviewers for the MIT Press. Marguerite Avery, ably aided by Katie Persons, waited patiently and shepherded skillfully. The manuscript was improved significantly by Paul Bethge.

To my children, family, and friends, thank you always. Ah, I love you guys!

The book is dedicated to Kristin Angela Thompson.

I STEM CELL BIOPOLITICS

ETHICAL CHOREOGRAPHY AT THE END OF THE
BEGINNING OF HUMAN PLURIPOTENT STEM CELL
RESEARCH

THE END OF THE BEGINNING

It is the morning of September 6, 2011, the second day of the seventh
session of the Intergovernmental Bioethics Committee's meeting at the
headquarters of UNESCO in Paris. It is ten years since President George
W. Bush delivered his famous speech on human embryonic stem cell
research from his ranch in Crawford, Texas; ten years since the twin towers
of the World Trade Center fell as I was interviewing students for my first
class about embryonic stem cell research. President Barack Obama has
rescinded Bush's stem cell policy, the news in the United States and in
Europe is about financial collapse, and the Occupy Wall Street movement
is taking its lead from the Arab Spring and the Spanish Indignants.[1]

The Intergovernmental Bioethics Committee's meeting is focused on
two topics. One is human cloning, including the moral status of the
embryos that have to be destroyed in research (that is, non-reproductive)
cloning and human pluripotent stem cell research[2]; the other is the role
and the scientific validity of traditional medicine. Both topics are framed
as searches for compatibility between faith and/or culture and modern
science, but the former is being treated as an issue primarily for the rich
industrialized countries and the latter as primarily of significance to the
so-called least developed or less developed countries. From the podium, a
member of the International Bioethics Committee (the expert committee
that reports on its work of the past year at the committee's annual meeting
of the Intergovernmental Bioethics Committee) wraps up the International
Bioethics Committee's reporting. In summing up, he reminds the delegates
that the International Bioethics Committee does not "produce a majority
report about anything," and that no consensus view has been reached on

the status of the embryo.[3] Disagreement is here to stay, he notes; whether you think embryos may be or should never be destroyed for scientific research, your view is "not a fault of reasoning." The very controversy that ignited debate a decade ago has become something upon which we must agree to disagree; it is time to move on to new bioethical topics.

During the question-and-comment period that follows, several of the delegates agree wholeheartedly. One delegate thinks that consensus on human cloning, including the research cloning under consideration, is nonetheless possible. Another reminds us that science has advanced so that embryos may soon no longer be needed, thus bypassing the embryo question. The delegates take up the task of making suggestions for the International Bioethics Committee's work for the coming year. The suggestions include several topics that are intimately tied to established stem cell research and its ethics: regenerative medicine; bio-banking; neuroscience; transplantation of and trafficking in organs, tissue, and cells; how trafficking divides the world into haves and have-nots; how to regulate bioethical issues in multicultural and multi-religious societies; "how to preserve the identity of the ones who give samples." The comments and suggestions reflect concerns specific to the countries and regions of origin of the delegates. They also underscore the shift away from the embryo and cloning and the permissibility of stem cell research and toward topics that begin with a world in which human pluripotent stem cell research and related technologies already exist, and in which transnational inequality, trafficking, and privacy will have to be governed carefully in the future.

This sentiment—that a tacit agreement to disagree has been achieved on the question of stem cell research with human embryos, and that there has been a partial turning of ethical attention elsewhere—signifies for me "the end of the beginning of human pluripotent stem cell research," and bookends the time period under consideration in this book. Manifest in many different ways of regulating stem cell research in countries pursuing the research, this state of affairs was attributable to political, ethical, and scientific work done over the course of more than ten years. It did not mean that disagreement about embryos, cloning, or anything else had gone away; co-existing as more or less active dissent alongside the normalization of the field, disagreement still tended to flare up when certain new technical capabilities or controversies surfaced.[4] It also did not mean that all nations or all factions within a country or a region had come to care about the same issues: I found evidence of national and regional specificity and a persistent pluralism of views, as well as national and international standardization and harmonization of the emergent field.

In this book, I explore what I call the "ethical choreography" of this process of consolidation of human embryonic stem cell research. I explore the geopolitics and the biopolitics of how, in California and beyond, some concerns brought about this normalization while other issues failed to gain much traction.[5] Understanding this ethical choreography throws light on what it would take to address important issues that are now relatively neglected.

During this period of consolidation, normalization, and public inuring amid controversy, stem cell research was an object of more multidisciplinary *ethical* debate and labor than is typical for advances in science and technology generally. For example, important work was achieved by this attention: attempts to "invent around" ethical roadblocks by scientists were transformative for the field, educational materials were developed, and access to therapies and to revenue were fought for. Many non-embryo-related ethical topics, as well as embryo-related politics, were brought into debate. Much of the debate was informed; some opinions changed; and regulatory, infrastructural, and scientific conditions for stem cell research were remarkably robust by the end of the period.[6]

I argue in this book that a high level of political attention to the ethics of the life sciences and biomedicine, like the attention given to stem cell research in this period, is a good thing for science and democracy. This argument is counter-intuitive for many, who fear—often with good reason—that non-scientific attention imposes unrealistic regulatory burdens on researchers, is anti-science (or critical of scientists), interferes with the relation of trust underlying the public funding of science, is a distraction to scientists and an invasion of research autonomy, stokes false hopes and fears about the field, misunderstands the role of science in stimulating the economy, and/or is under-informed about the science. If attention from non-scientists in fact slowed down research aimed at finding cures and undermined freedom of inquiry and the public understanding of science, such attention could hardly claim to serve democracy or science. It is tempting to think that it would be better to hone the system of regulatory review and ethics to be as lean, automatic, and smoothly functioning as is possible.

The progress of science itself also might make it seem that less, not more, ethical attention is now required than when the field was new. Human pluripotent stem cell research made for controversy and garnered attention at first because it needed human embryos as its raw materials, and many people felt that it was unacceptable to use embryos as research substrates. At the same time, the ability to obtain cells that had the potential to turn into any of the body's specialized cells promised both the ability to replace old or faulty cells with new cells and the ability to make these repairs using cells from the patient's own body—a scientific and medical revolution. Less than ten years later, it was possible to induce pluripotency in certain somatic (non-embryonic) cells, without using embryos. Scientists also worked out how to deliver the "immortalizing" transcription factor genes without viral vectors, so that the cells had the potential to be used therapeutically in transplantation and other medical applications. As I review this introductory chapter, rewiring the fate of cells, without having to go through the intermediary step of reverse programming them back to induced pluripotency, is also being pursued, and the first successful animal studies of "spell-checking" (correcting a faulty gene in an induced pluripotent stem cell line) has pointed the way to a future in which gene therapy, bio-informatics, and stem cell research may work in tandem.[7] Some continue to argue for the superiority of stem cell research involving embryos, especially in the light of promising cloning results; others seem to have decided that the controversy around embryos is moot at this point because embryos are no longer needed.[8] Progress in rewiring cells' fates may mean that particular starting and ending points for cells, rather than pluripotency, will become the focus. In this case, pluripotency may already be on its way to becoming an important liminal concept, a proof of principle and standard of efficacy, rather than a state perilously close to the embryo through which all regenerative medicine must take its cells.

If procuring human gametes and embryos is less necessary to the field, and if the debate hovering around the life potential of embryo-like pluripotent states is settled (without being agreed upon), one might argue, on scientific grounds and procedural grounds, that intense ethical scrutiny would only undermine this stability. What, then, are the grounds for calling for renewed rather than diminished multidisciplinary ethical attention to the field of stem cell research? The answer is threefold: There is still work to be done on international and national standards for cloning and embryo and gamete recruitment, as these issues do not go away. There is an urgent need to turn to ethical issues that were largely put aside during the first phase of debate, such as health disparities, military applications of stem cell research, and who decides which conditions need treating and what constitutes a cure. And there is a growing need to bring attention to ethical issues that are arising as the field becomes more integrated into biomedical research and life-sciences research in general. Stem cell research is an integral part of a convergence of bio-informatic data mining, synthetic biology, neuroscience, and regenerative medicine that is well underway.[9] Much like the information revolution of the past 25 years, this "personalized medicine" revolution is likely to change almost everything in the life sciences and in biomedicine, and in ways we cannot predict. We are entering a phase of history in which enhancement may be routine for some people, repairing many kinds of human tissue may become practicable, life may be significantly extended and enhanced for some, and medicine and cures may become customized. At the same time, it seems likely that the majority of the world's people will benefit little if at all from these developments, and the less privileged and the least privileged will continue to be over-sampled as research subjects and substrates.[10]

My job in this book is to make a case that, where multidisciplinary ethical attention is concerned, promoting good science and good science's concomitant, meaningful ethics go hand in hand. In part I of the book, I explore the initial attention garnered by the embryo debate that spilled out into all kinds of innovative ethical thinking but also quickly coalesced around a few nodes of regulatory procedure. In part II, I take up the kind of ethical inquiry that might form part of renewed public and private engagement with stem cell research as it matures and integrates with other innovative fields. As an alternative to the insufficiently participatory emphasis on human dignity characteristic of the Bush stem cell era, or the excessive reliance on rationality and process of the deliberative ethics of the Obama era, I argue for an ethical choreography of vital ethics.[11] I seek good science in which knowledge and ethics are held to the highest standards; for this, ethics must live, as much as innovation.

NARRATIVE TRAJECTORY

This is a book about the normalization of stem cell research, beginning with the first successful derivation and immortalization of human embry-

onic stem cell lines at the end of the twentieth century[12] and, in the
United States, President George W. Bush's stem cell policy of 2001. A few
years later, several individual states enacted stem cell funding and infrastructure initiatives. Things quickly changed again with the development
of the capacity to induce pluripotency in stem cells without using eggs
or embryos as raw materials,[13] quieting the embryo debate somewhat. In
the United States, the end of the beginning began with President Obama's
2009 repeal of the Bush policy and concluded gradually during its legal
aftermath through 2012. In the meantime, California passed Proposition
71 (the Stem Cell Research and Cures Act) and several nations competed
for scientific priority.[14]

In this period, debate about human embryonic stem cell research in the
United States centered on national political battles, and comparable battles
in individual states, over whether or not the embryo-using research should
be funded by the government, with opponents of abortion pitted against
pro-research and pro-cures factions. Competing conceptions of the future
were also evident in and around the debate, staking out a different, non-
embryo-centric basis of dissent. At least in California, the latter operated
primarily in the realm of rhetorical framing of the future. Some, dubbed
"science pushers" and accused of hype by the skeptical, imagined a future
including customized cures, economic rejuvenation, and life enhancement
and extension; they lamented what they perceived as a regressive anti-
science tradition in American culture and politics, and saw themselves as
spokespeople for rationality, truth, innovation, the health of the U.S.
economy, and progress. Others, self-identified as progressive but dubbed
technophobic and paternalistic by their detractors, sensed risks in proceeding with these technologies; they evidenced a less trusting attitude to
science and its associated economies, and expressed fears that a lack of
regulation would lead to breaches of privacy and humanity and would risk
bringing us a world in which enhanced "haves" would select out or vivisect
a local and global underclass of "have-nots."[15] Many people, perhaps most,
felt a mixture of concerns about the field and excitement about its scientific and medical promises. These two debates—the partisan U.S. political
debate in which human embryos stood in for Republican and Democratic
world views about abortion, the role of government, religion, science, the
economy, and values and the civic debate about utopian versus dystopian
visions of the future—pervaded the science, the ethics, and the politics of

regenerative medicine in California and beyond during the period with which this book is concerned.

TRIAGE: ACTORS, FIELD SITES, TRANSCRIPTS

I work at the intersections of science and technology studies (a field that combines sociology, anthropology, history, and philosophy, of the sciences, technology, information, and medicine) and transnational feminisms, many strands of which inform my approach to the emergence of stem cell research. In particular, this project is influenced by and is in conversation with the work of scholars who have documented and theorized the ongoing redefinition of life, death, reproduction, and health and their relations to governance in the age of biomedicine.[16] Because of stem cell research's positioning as up for public debate as a condition of its possibility, and because of roles I played during the research, this book takes a methodological and theoretical turn toward a more normative, policy-relevant approach to analyzing science and technology in society.[17] My research centered on the construction of and the disagreements over the future of stem cell research, which rested on linking possible futures to particular pasts and presents and not others.[18] This, in turn, led me to argue for particular futures about which I came to care a great deal, and to attempt to link such possible futures to pasts and presents that could give them political legitimacy and saliency.[19]

In my book *Making Parents: The Ontological Choreography of Reproductive Technologies*, I worked primarily ethnographically. Many of the sites at which I worked in the course of my project on stem cell research and related technologies are shared with my earlier work on reproductive technologies; indeed, I began studying stem cell research because eggs and embryos were raw materials for the field. My previous concern with the disposition, circulation, and capitalization of *ex vivo* embryos and gametes has continued with stem cell research. At the same time, the fields of human embryonic stem cell research and reproductive technologies quickly proved surprisingly different, despite enduring links; the research, researchers, research subjects, and research beneficiaries, promoters, and naysayers of one field are not the same and are not treated in the same way as those of the other field.[20]

I began the research for this book by following and participating in these debates with an ethnographer's and participant observer's orientation to the "thick present." What was happening in the present included work by a large number of people, and expenditures of time, money, institutional capacity, and other resources, to graft stem cell research onto relevant pasts and to intimate various futures. Stem cell research had to be steered away from the politics of abortion to be funded and normalized, but also away from under-regulated and scandal-riven aspects of reproductive technologies (especially the private eugenic market in gametes) and gene therapy, despite the material, scientific, medical, ethical, political and institutional proximity of these fields to human pluripotent stem cell research.[21] When I began the project, because I was an ethnographer, I filed for Human Subjects permission to carry out interviews and ethnography. This required that I specify my subject populations and their degree of vulnerability, as well as my sites and my recruitment efforts. I quickly put that protocol aside, however.[22] I wanted data from sites and subjects whose identities were appearing only as events unfolded. The people I originally intended to interview came to mind because they had already taken positions, or were positioned by their medical condition, vis-à-vis stem cell research, and in many cases because they were interested in persuading me and others to take their positions. Something was missing in this way of proceeding. For example, the "general public" (a terrible expression masking all kinds) was integral to the story about the making of the "populist" California stem cell initiative known as Proposition 71 and its comfortable majority, yet when I began working on stem cell education "members of the public" in California were routinely cast as knowing too little about the science of stem cell research. "Public engagement" tended to mean subjecting various audiences to stem cell education efforts by those representing institutions carrying out research. How could stem cell research hinge on the democratic wisdom of the people *and* widespread ignorance be imputed to the public at the same time? Similarly, my institutional-review-board-approved questions ("How do you feel about egg donation for stem cell research?" "Do you think it is acceptable to use a leftover embryo from an IVF clinic for stem cell research?") began to seem as if they had been drawn from someone else's project. Meanwhile, I found my project focusing more on terrible interview questions (terrible because I didn't want to know how an individual would answer them but instead wanted to follow the questions themselves) such as "Why is California acting like a nation?" "Why does no ethical work happen in an ethics committee?" and "How can you ask the public to pay for and then not guarantee that they will benefit from stem cell therapies?" and on answers that people I had not solicited for interviews gave to my erstwhile interview questions. I increasingly found that I wanted to begin to capture and tease apart what were—from my perspective—evolving, fractured, competing, animating scripts of the field, not the scripts that structured the dominant debate.

If reproductive technologies have always played out in ways that throw light on political debates of the day, human embryonic stem cell research actually *was* political debate of the day in the United States. On the right, it was always political because of the abortion issue. On the left, various parties—some strongly allied with the Democrats and the Obama administration and its regulatory and bioethical bodies, others prominent in progressive civil-society groups—sought actively during the period under consideration in this book to present a bioethical left that was also part of political culture and debate.[23] Stem cell research in the United States was from the start a salvo in national and international political debates about innovation, abortion, and competition in ways that were over-determined and under-situated, with bioethics as a lingua franca zone of contestation. This politicized aspect of stem cell research seemed to require, at the very least, that I add a regulatory archive to my ethnographic one. Drafts of policy and legal documents, for example, were as relevant as my site-based field notes, and participating in ethics and policy was as revealing as participant observation (and was, in any case, asked of me). The pervasiveness of ethics also gave the field a non-optionally normative cast, tying the delocalized site and the more normative focus of inquiry together.[24]

The task of compiling and processing my archive required new practices to augment the fieldwork that had previously formed the bedrock of my technical practice. To analyze this episode of scientific and medical history in the making in the way I wanted, I developed a method that I came to call "triage." "Triage" is a word associated with waiting to be seen in the emergency room of a hospital.[25] It is the process by which those waiting for treatment of more serious and less serious conditions are called up to be seen by doctors and nurses, and it is greatly affected in the United States by the large numbers of uninsured people who are able to receive medical care only in emergency rooms, many of whose conditions are acute because of under-treatment and/or structural inequality. Historically, and in most dictionaries, triage is defined as prioritizing the sick and the wounded—usually on the basis of some combination of urgency, treatability of the injury, feasibility, and safety of transport where required, and the value of the life to be saved—where there are insufficient resources for all to be treated immediately and fully. In most countries triage has been codified during times of war and other medical emergencies. During my research, triage (especially with its U.S. connotations of the medically underserved) came to seem a less and less metaphorical, and more and more empirical, description of the processes by which some issues (lives) came to the fore as stem cell research was normalized and other issues were left in the waiting room (i.e., to be dealt with later), or on the battlefield to be counted among the dead (i.e., forming the necropolitical side of the biopolitics of stem cell research).[26] "Triage" described my process of selectively acquiring data and building my archive. In an environment of an excess of data (documents, field sites, actors), all resonating in a political debate in which local, state, national, and international topographies were at stake in intersecting ways, I prioritized some actors, sites, and transcripts over others according to inquiries about different kinds of lives and values. This methodological triage concerned my three major foci for data collection, and structured my archive: actors, including their often multiple institutional locations; sites in the story, more and less localized; and the various transcripts, some historically significant and some not at all so, some already directing practice and some under preparation, that echoed through and structured the field site across time and space.

ACTOR TRIAGE: FIVE FIGURES

In the course of several years' work in and around stem cell policy making, thinking, teaching, and event hosting, a number of composite figures emerged and critically informed the writing of this book. The sketches below, drawn from my field notes, illustrate these figures, who as much as any individual scientist's or ethicist's work, or any institution's role or stance, defined, defended, and critiqued the field as it unfolded, from my perspective. These five figures—the engaged student who probably will be conscripted body and soul into personalized medicine in her or his lifetime; women debating how human embryos and eggs do and don't fit into gift and therapeutic and commercial and research trajectories for body tissue; patients showing border-defying impatience; various kinds of scientists giving voice to the sheer (and non-innocent) thrill and work of the science and innovation involved; and those living with and thinking about non-normative kinds of embodiment who question what cures will mean and who gets to decide—were the triage actors in my field site.

Each of the comments clustered below in five figures came from an individual, each with her or his own expertise, desires, experience, hopes, and fears that are in no way captured by these figures; the figures do not "speak for," represent, or contain any individual.[27] Neither do these figures simply reflect particular political or demographic characteristics.[28] Rather, these figures give structure to voices I came to carry around with me—a cast of composite, challenging, worthy interlocutors. The figures are based on the following:

• the comments and questions of ten different (ranging from undergraduate to graduate, identifying as of different race, gender, sexuality, immigrant status, or major field of study) engaged students—prime targets of gamete donation advertising and genomics testing on U.S. campuses—on cell and tissue donation
• reflections by eight women varying in age, race, class, religion, country of residence on egg or embryo donation for stem cell research
• reflections by two male patients (one in the United States and one elsewhere) interested in stem cell research
• reflections by a computer scientist, two life scientists, and a student engineer (three men, and one woman, three in the United States, and one elsewhere) on the promise of regenerative medicine to the sciences of the future
• reflections of four individuals concerned with the relation between stem cell research and disability (two students, one activist, and one professor). Each cluster carries with it different framings of the issues at stake, and each profoundly shaped (and sometimes troubled) my sensibilities to the field.[29]

FIGURE I   A student who was conceived with the help of a sperm donor discusses with me the seemingly inexorable movement, aided greatly by social network technologies, to find one's biological parents and half-siblings, and the associated trend toward known donors; she asks whether children who were adopted as embryos may join these networks. Another student asks "If there was a reunion for donor children, would the Snowflakes come, and would they get on with all the queer families using donor gametes and surrogates?" On another occasion, two students raise issues about privacy, anonymity, and reciprocity: "If gamete donors should be known, should embryo and stem cell and tissue donors be known too?" "If donors are known, do they have any financial responsibility, such as paying for children?" On yet another occasion we discuss who gets intellectual property from known donor cells: "Do donors have any chance of getting rich if their cells lead to big scientific breakthroughs?" I am asked what I think about the egg-trafficking rings that have been in the news recently. An undergraduate links the economy to putting one's body to research work: "I could see why people want to be in clinical trials or to donate eggs or cells or whatever: you can get some money in a bad economy, it's work." "For women you can take care of your children while you do it; you might even get health care while you're doing it or get to emigrate to a richer country if the scientists or rich patients want you," another student elaborates. On the question of the pros and cons of putting new biological capabilities in the hands of consumers ("direct-to-consumer," i.e., mediating access to biomedical tests through a retail outlet rather than a physician or other expert), a student suggests that "if Walmart would make cheap stem cell kits, we could patent everything unique about our own cells, and then companies would have to buy them from us." "Just imagine," the student goes on, "my cell line could make you immune to the next flu epidemic, and I get a fat check for my twenty-first birthday!"

FIGURE 2   A young woman talks about donating her eggs for stem cell research versus for *in vitro* fertilization (IVF). She prefers to donate for that purpose, as she believes she will be treated better, will be paid more, and will have done a genuinely generous thing. Another young woman, who hasn't yet had a child, says that if she were to donate her eggs she might ask the doctors to freeze some eggs for her in case she needed them later, and donate half of them only to the couple. "Can you do that?" she asks. Should the need ever arise for her to seek IVF to get pregnant herself, she says, she would donate any leftover embryos from her own treatment to research, though. She would donate to research because she wouldn't want someone else to have her child if she didn't get pregnant herself. She also wouldn't want there to be half-siblings to her own children from the same batch of eggs, the same age, but not known to one another: "What if they met in high school and fell in love? Or what if I had only one child and he or she desperately wanted a sibling, and the sibling existed but we didn't know about it? Or we did know about it, but we couldn't contact them?" An older woman who went through IVF tells me that ovarian stimulation would be a lot to go through for research alone. "It would be weird to get research materials right out of my body, in a risky procedure, and I wouldn't feel as good because it wouldn't be helping anyone at the emotional level," says a woman who had been paid for being a surrogate in the past. "There isn't really any connection between donating eggs and research, but helping someone get pregnant makes sense, it's what the body's doing anyway," suggests another. One young woman sees the money from donating eggs to research as a straightforward way to earn some money with which to finish her education and take care of her young child without the added emotional difficulties of having been an egg donor or surrogate for a pregnancy and a future living child. A middle-aged woman wishes she had had the chance to donate eggs to research while she still could. "The science," she says, "is very exciting." Another middle-aged woman sees egg donation for research as part of a much larger trend toward commodification of the bodies of women, especially low-income and minority women in the United States, and trafficking in women within and across national borders, and believes the practice should be criminalized. In any case, she thinks that stem cell research is a bad use of public health-care funds, and she doesn't see why the public should foot the bill for research that isn't likely to benefit those without medical insurance.

FIGURE 3   An old man with stage 4 cancer writes that to stay alive he will do anything that medicine makes possible, even though he is a devout Catholic. He worries about the burden that caring for him places on his family, but not too much: "You have to fight when your life is at stake." He is grateful to live in a country with generous nationalized health care. He is disdainful of the lack of universal health care in the United States despite its relative wealth. But he is more angry about the stance of the United States toward Palestine, which he thinks is holding up the progress of history: "It's all connected: peace in the Middle East; our dependence on oil; wasting our money on wars instead of cures." A young man says he would go to China for a stem cell treatment if he could. He doesn't see why everyone is demonizing countries that are offering treatment even if it is still experimental. "In the United States," he says, "people will die waiting for the FDA."

FIGURE 4   A middle-aged computer scientist is invigorated by the coming confluence of neuroscience, regenerative medicine, bio-informatics, and bioengineering. The idea of being enhanced and of living longer seems self-evidently the greatest challenge and source of wonderment for the twenty-first century. A scientist interested in the age-defying potential of stem cell research provokes his audience by claiming that global health disparities would not necessarily be exacerbated by increasing longevity, and that the population pyramid, while changing shape because of the increase of older people, would not necessarily be a problem: "Old people would no longer be old, that's the point." An engineering student thinks that at present we can support the world's population because life expectancy is still low in poor countries, but eventually we will have to face the global consequences of the abilities to cure disease and prolong life that stem cell research and personalized medicine promise. She is not sure if we are tying up our inventions with intellectual property and trade secrets, keeping cures and technical infrastructure out of the hands of the poor, or if technology transfer is possible only if there is a strong international intellectual-property regime. Another scientist tells me there is no reason not to develop these technologies in a radically open-source manner. "Knowledge wants to be free," he says. "But freedom also means research autonomy, and stem cell critics don't understand the science and are holding up research."

FIGURE 5   A woman expresses her commitment to the social model of disability, rather than the medical model, describing her body as disabled relative to a norm and as in an environment designed around so-called able-bodied people, rather than a body "without ability." A student talks about her interactions with medicine. She wants to be able to go to good doctors and get help for the specific ailments from which she suffers. "Disability rights and wanting good doctors are not incompatible," she says. She also wants to define for herself—perhaps changing her views over time, in conversation with others whose views she respects, and in the light of what is technically possible—how she feels about "curing" her condition, whatever that might mean. She doesn't think able-bodied or healthy people should decide what needs curing. "Nothing about us, without us," she recites. It troubles another student that some prominent advocates of stem cell research are not the people with the disease or condition in question but parents of children or young adults who have the disease or condition and probably could speak for themselves. "It is almost as if the parent could not accept their children less than perfect," she comments, although she sees that parents suffer, too, because there aren't enough resources, and fear for their children. She points out that disability doesn't mean a suffering life: "People should get over that. It's a life with joy and suffering like any other." An academic is deeply suspicious of prenatal and postnatal selective technologies and enhancements and does not understand why so few people recognize that eugenics never went away. "There is some humor in the fact that soon no one will be good enough," this scholar jokes. "We're all in need of enhancement now!"

FIELD-SITE TRIAGE   At the beginning of the period covered in this book, human embryonic stem cell research was not yet clinical (only hematopoietic stem cell research had a proven clinical record), although it came with a promissory clinical narrative of cures. Laboratories doing stem cell research were springing up or being retooled in the public and private sectors at every phase of the proverbial arc from bench to bedside. On campuses such as my own there were labs in bioengineering, neuroscience, immunology, molecular and cell biology, and clinical research, among others, that seemed institutionally and epistemologically, as well as in terms of their desired outcomes of research, very different one from another.[30] I visited labs and clinical environments and attended scientific (as well as ethical and policy) meetings, and ethnographic research in these spaces was vital to my research. Nonetheless, it was the more or less public, diffuse rather than situated, conversation about stem cell research that continued to demand my attention. Whether looking at scientific and technical practice, at bureaucratic forms, at political rhetoric, at legal or economic concerns, or at any of the other elements that helped define the field, the location of sites and links among them were less fixed than I had at first imagined.[31]

During the period in which international competitiveness became one of the issues that framed federal and state arguments for stem cell funding in the United States (roughly the years 2004–2008), I visited a number of labs and talked to policy makers, scientists, and activists in some of the countries in question. As I discuss in chapter 3, my conventional ethnographic exposure was limited by time, language, and funding. These very limitations, however, meant that I focused on the international aspects of the national controversy (and vice versa), for although I could see how the labs differed from how they had been portrayed at home by the U.S. and international media, I was not at any point tempted to think that I had understood what was "really going on." From a transnational feminist perspective, neither the international nor the national spaces of science are "flat" (without hierarchy); instead, the national and international worlds of science have complex, overlapping topographies that include the history and the politics of scientific knowledge.[32]

TRANSCRIPT TRIAGE

The documentation about stem cell research includes scientific papers and editorials, textbooks on technical practice and on medical ethics, journalistic accounts, corporate reports, advertising copy, regulatory documentation of animal or human research, humanistic and social-scientific books about the meanings of the research, policy documents for all kinds of institutions and jurisdictions, legal briefs and opinions, political bills, position papers written by non-governmental organizations or religious groups or special interest lobbies, scientists' and patients' memoirs, and more.[33] Documentation also is usually not hard to access. Nonetheless, some of it is still, in a significant sense, invisible: the sheer volume of written materials precludes paying attention to everything, some documents get much more attention than others within the field, and things not written down can be extremely important.[34] In fields that are unfolding and not yet stable, as was true of stem cell research in the period I am considering, the activities and principles to which documents give life have yet to settle.[35] Although an enormous amount of documentation related to stem cell research is freely available and searchable online, some is on the "invisible web," hidden within proprietary databases or behind monetary or security barriers, introducing its own paradoxes of classification, storage, and searching.[36]

The appendixes present text from, or URLs for, the documents that, through the process of triage, most centrally inform the chapters. Some—

such as the texts of presidential speeches on stem cell policy and the text of Proposition 71—form primary evidentiary bases for a chapter, and I do a close reading of them in the text. Others—including the Common Rule and the Nuremburg Code—had the status of shared background in the field, and they were referenced in new ethical documents. Others were documents whose preparation was ongoing, such as amendments to California's constitution. Yet others were integral to ethical review, such as Stem Cell Research Oversight forms. They are all primary documents, but their functions differed; the appendixes reflect my own archival interests and my emerging classificatory interests.[37] I hope that this triaged stem cell archive will prove a helpful resource, allowing easy access and reference to the texts in question.

OVERVIEW OF THE BOOK

Part I consists of two chapters that contribute to the ever-growing literature on the politics, economics, and ethics of the life sciences. In this introductory chapter, I have introduced the narrative arc of the book and explained the method of triage that is used. In chapter 2, I explore stem cell research as a science that is widely considered to engage ethical concerns. I investigate the logic of promoting stem cell innovation, which during the period under study involved a pro-cure aspiration and a focus of ethics around the question of how to procure the cells and cell lines for doing human pluripotent stem cell research. I contrast this pro-cures and procurement logic with the kinds of ethical concerns that come into view if stem cell research is conceptualized not just as innovation but also as part of the overall picture of health care and as a recipient of significant public funding. I argue that the ethical choreography needed now that we have reached the end of the beginning of human pluripotent stem cell research would situate innovation in this larger framing and would see pro-cures and procurement logic as just one way through this science. Finally, I ask how sciences with ethics should be governed, and what their role is in governance, especially in democracies. I consider the dignity ethics (resonant with a pro-life philosophy) of the Bush era and the deliberative ethics (resonant with deliberative democracy) of the Obama era, and argue that human pluripotent stem cell ethics can and should go beyond these two poles.

Part II is also made up of two chapters, one exploring the domestic situation as it ricocheted from the federal level to the state level and back again and one following transnational circuits that animated and were in turn animated by the stem cell politics of the United States and other countries. In chapter 3, I consider the three phases that together made up the beginning of human pluripotent stem cell research in the U.S: the time around President Bush's 2001 national policy; the period in which states "seceded" from national policy, during which California's Proposition 71 pioneered state-led stem cell research; and the period around President Obama's 2009 policy. I compare and contrast the two presidents' speeches and their relation to policy making, then describe what it was like working in the immediate aftermath of Proposition 71, as policy was being formed. I consider the relevance of California's initiative for women through a discussion of the procurement of eggs and embryos, and the relation of pro-cures rhetoric to disability rights and to bench-to-bedside translational models of research. I also look at participation in and benefit from stem cell research, and stem cell research's uneasy relation to the public (especially the medically underserved). I contrast the debate about women and egg donation with what I call the "real estate" framing of the proposition itself. In chapter 4, I turn to the transnational geopolitics of stem cell research, starting from a California perspective, during the time in question. Advocates for stem cell research raised concerns about international competitiveness throughout the period covered in this book, highlighting what they saw as the dampening effect of restrictive legislation around human embryos on U.S. innovation. During the campaign for Proposition 71, California was posited as an oasis for stem cell funding and innovation, solving the problem of the absence of federal funding that was precipitating a brain drain of stem cell scientists to other countries with more stem-cell-research-friendly working conditions. In the early days of the California Institute for Regenerative Medicine, putative developments in stem cell research itself, as well as in stem cell rhetoric, began to focus on South Korea, Singapore, and Britain as magnets for the best and the brightest. The difference between the regulatory climate of the two Asian countries and that of the United States was sometimes caricatured as the difference between an "East" with a pro-science-and-technology spirit and a lack of concern for the moral status of the embryo and a "West" that had been turned away from its traditional scientific superiority by a minority of anti-science religious fanatics worrying more about embryos than about the economy and cures, and an even smaller minority of technophobic progressives worried about the integrity of nature and critical toward science. Whatever the proverbial grains of truth in this caricature, it presented the world of stem cell research as if how it appeared from California was also how it appeared from other places.

I traveled to Singapore and South Korea (and to Britain and Thailand), visited labs and facilities, and met with ethicists and scientists. The "East" that I was hearing about in America as the foil to fears about international competitiveness matched poorly with the *in situ* realities of stem cell research in Singapore and South Korea. Perhaps not surprisingly, those two countries and their flagship facilities of the time not only differed from this image of the East; they also differed markedly from one another, and people I met in both places discussed their own nationalist, regional, and international aspirations, just as their counterparts in the United States and particularly in the state of California did.

Toward the end of the period under discussion in this book, Asia was again in the news from a U.S. perspective, but no longer as a brain-drain threat. Stem cell research was becoming increasingly international, and formal collaborations were being announced, notably by the California Institute for Regenerative Medicine. For convenience, I date this next phase to the respective announcements by Shinya Yamanaka at Kyoto University (after a stint in California) and James Thomson at the University of Wisconsin that somatic cells could be reprogrammed to form induced pluripotent stem cells. At the same time, the question of transnational stem cell tourism began to be taken seriously. Rather than the "Asian Tigers" of five years earlier, or the Japan or Singapore or any of many other countries in Asia making rapid strides in stem cell research, India and China were targeted as hubs for unregulated stem cell tourism. Stem cell researchers from around the world, including Asia, worked under the auspices of the International Society for Stem Cell Research to draw a line between places offering "snake-oil" cures, and luring desperate patients with unsubstantiated claims for dangerous stem cell procedures, and genuine if experimental stem cell research. The patient advocate, prized in U.S. stem cell research as the embodiment of self-willed agency, appeared to have little in common with the victimized medical tourist now being protected. One provided the moral case for a nascent, controversial field, and continued to speak for the field; the other helped the field to draw boundaries around itself and to cement its international legitimacy and maturity by excluding "bad science" practiced elsewhere.

Part III moves the inquiry from the geopolitics of stem cell research to the no-less-political questions of research subjecthood. Apart from patients, scientists, activists, and innovators, the first 15 years of pluripotent stem cell research involved human subjects as donors (to themselves and/or to others) of gametes and other cells for research. Stem cell research also used human embryos, various kinds of pluripotent cells, and tissue from fetuses, which might have been, but only sometimes were, considered to entrain additional subjecthood. Pluripotent stem cell research also used animals in ways that were integral to how good science and good ethical practice were understood and regulated in the field. In part III, I investigate the logics of research subjecthood and consider challenges posed to pre-existing practice posed by the promise of regenerative medicine. In chapter 5, I consider how, at critical moments, norms of altruism, anonymity, and the alienation of tissue from donors all came under renewed scrutiny, and were increasingly accompanied by demands for various forms of reciprocity. Challenges to anonymity arising from forensic and bio-informatic developments and challenges to altruism and alienation from the value chain after donation, as well as from distributive justice perspectives, converged to suggest that the old model for donation was no longer working. Although the advent of induced pluripotent stem cell lines removed many of the field's problems associated with procurement and provenance, it also allowed cell and tissue donation to slip back into older patterns of the donation, the use, the buying, and the selling of human tissue. This threatened to undo the ethical work that had been done on donation and reciprocity at exactly the moment when such work became more and more necessary. In chapter 6, I investigate what I call the substitutive research subject. I argue that regenerative medicine's promise of the autologous cure (involving healthy cells, derived from one's own stem cells, that one's body will not reject) translates to an ethical promise to de-conscript the post–World War II substitutive research subject. I also examine the logic of substitution of non-human animals for humans. In particular, I trace the ethics of substitution of animals for humans as it was reiterated in my field sites and in the documentary genealogies used in these sites. The epistemological and regulatory embedding of non-human animal substitution is deep; at the same time, there is widespread agreement that stem cell research done in animal models may not work well for humans. Cell therapies, some argue, behave differently in different species; likewise, for stem cells to deliver their full therapeutic promise, they should come from the patient being treated, rather than from a conventional model-and-test system of clinical development. Alongside efforts to further humanize and better visualize animal models, there are some calls to move out of animal systems and into *in vitro* systems that combine tissue engineering, synthetic biology, and stem cell research, but not enough yet to make it a concerted goal. To give traction to the ethical arguments, as well as scientific ones, for moving away from the substitutive research subject, I propose the idea of the "greater moral universe." Unlike ethical positions that attribute subjectivity based on rationality and/or empathy, I make a *ceteris paribus* case for attributing subjecthood even when shared rationality cannot be assumed, and even when one is unable to imagine oneself in a similar position.

SCIENCES THAT "HAVE ETHICS"

Human embryonic stem cell research "has ethics" in a strong sense. That is, just about everyone who encounters stem cell research takes it for granted that it is enmeshed in ethical controversy, especially in the United States. To talk about stem cell research is to talk about ethics—but what is ethics, that most ancient of inquiries, in this context? Sciences that "have ethics" in the sense that is of interest to me have ethics as part of a layer of disciplinary and bureaucratic activity increasingly attached to the bio-medical and human life sciences, and scathingly dubbed the "ethics indus-try."[1] The conjunction of the words "ethics" and "industry" reflects the profoundly capitalized, or capital-aspiring, aspects of the biomedical and life sciences, especially in the United States, with professional ethicists deployed to negotiate emerging regulatory constraints. The conflation of ethics with industry, while telling, is too reductive, however. One can "follow the money" to get at "the real story" of stem cell research, but the money one is following cannot yet flow (as if) independently of the ethical, regulatory, and material cultural conditions for creating markets in and around human pluripotent stem cells.[2] Some markets in the biomedi-cal and life sciences, such as pharmaceuticals markets, are deeply entrenched parts of the capitalist world order.[3] Other markets whose raw materials or testing substrates are human bodies and body parts, and whose empirical effectiveness has not yet been demonstrated and cannot be demonstrated without first using bodies and body parts, cannot simply pronounce or finance themselves into being established markets. Markets in biomedical and life sciences are constantly up against a triple bind: difficulty in con-ferring public or private ownership (intellectual or other) upon, and generating profit from, that which is natural; difficulty in commodifying, buying, and selling persons and their body parts; and difficulty in applying differentiated access to and market pricing of care services, including health care.[4] For established markets in the biomedical and human life sciences, these sources of tension may be relegated to skirmishes at the borders; for newer fields, including stem cell research, following the money means, at least in part, following the ethics.[5]

I use the word "ethics" in this book to refer to the wide-ranging activities, including formal bioethical policy making, in which various actors engaged during my research (myself included) to advocate for some ways of proceeding with pluripotent stem cell research over others on the grounds that they would be better for some people or things in some way. As such, ethics is an overarching normative term for me, ranging in its application from political contests over funding, rhetoric, and institution building to matters of personal belief and normative arguments made by scholars and activists hailing from a range of disciplines and social locations.[6] In this chapter, I am concerned with the dominant ethical landscape—which, based on three of its most salient characteristics (see below), I refer to as "pro-curial"—that emerged during the first phase of human embryonic stem cell research in the United States, and in California in particular. I focus on what pasts, presents, and futures it activates and is activated by, and what the consequences of that are for various constituencies. During the presidency of George W. Bush, architects and promoters of the pro-curial frame emphasized its contrast with the ethics of human dignity. During the presidency of Barack Obama, a kind of deliberative democratic bioethics became the hallmark of efforts to make the national bioethical process compatible with the innovation aspirations of the pro-curial frame. Despite the promise of combining ethics with innovation, however, many barriers to biopolitical inclusion in the pro-curial frame remained unaddressed. Likewise, democratic bioethics, resting on an implicit ideal of consensus through reason, provided no tools for taking persistent dissent into account or for dealing with challenges to the limits of reason in bioethics. At the end of the beginning of hESC research in the United States, the research had become possible, and even normalized as scientific and ethical practice in many states, even though the original struggle between pro-life and pro-stem-cell-research advocates had not disappeared, and alternative bioethical issues continued to struggle to find voice.

Sciences with ethics, because of the attention given to their conduct, offer an unprecedented opportunity to reset how science as usual is carried out. The end of the beginning of stem cell research must open up, not close down, what can be raised as ethically important in the field. This will only become more important over time because of stem cell research's links to other emerging or rapidly growing fields. The hESC-related human biomedical fields of genomics and bio-informatics, personalized and regenerative medicine, reproductive and pre-natal selective technologies, human enhancement and longevity sciences, neurosciences, and synthetic biology could also be said to "have ethics", because they intersect with controversial topics (eugenics, genetic privacy, commercialization, regeneration, and/or weaponization of human life). These rapidly growing fields are converging to bring the human body and our environments into a new "post-human" era; ethics is integral to them at every phase, from how their raw materials get to the lab bench all the way to the patient's bedside or the marketplace and beyond, and the convergence of these fields is only likely to intensify the urgency of ethical concerns.[7] The ethics of sciences that have ethics can be contrasted with a more conventional view of the ethics of science made up of a professional code of conduct (don't fake your results, don't steal my reagent) and possible downstream ethical, legal, or social implications (after the science is over, are the results used for good or ill?). It may well be the case that *in fact* (and certainly in fiction) most sciences and technologies have always "had ethics" through and through— it is, after all, a hallmark of many sciences that their raw materials, whether living subjects or extracted natural resources, cannot be procured without complex social organization that implicitly and explicitly involves competing ethical commitments. But most sciences are not treated as if they were always already embroiled in ethical controversy, and controversy that does exist is not typically shorthand for the views of competing political parties or other debates that get the attention of a large constituency on the science in question, in the way that it is for stem cell research.[8]

It has taken a great deal of work from scholars of the social study of science over several decades to argue for and illustrate the extent to which natural and social orders are co-produced.[9] For sciences that have ethics in the sense I am referring to in this chapter, no one has to be persuaded that they have ethics; meticulous scholarship, certainly, is not needed to convince readers that science and ethics interact in these fields of research.

Instead, it is those who want to act as if the science were "just science" who have the hard case to make, and who have to rely on the work of ethicists, legislators, and rhetoricians, often drawn from the ranks of those with an economic or personal interest in the outcomes of the research, to carve out spaces where they can get on with their work. What can be learned from the challenge of sciences that have ethics, in terms of social and moral theory, and what are the implications for governance in societies that have more and more sciences that have ethics? "Good science," in this book, connotes the conduct of sciences that have ethics in ways that iteratively develop the science and ethics of their fields together to the mutually entwined and multiple ends of both robust science and technology, and the greater articulation and mitigation of problems of distributive or other injustice. This ethical choreography is necessary if we are to govern in such a way as to lead to mutual possibility of vital science and vital ethics.[10]

The social problems of many sciences and technologies stem from their having proceeded, in the main, according to the tacit consent of our institutions and regulations and division of labor in the academy and beyond, as if there were no ethical, social, or legal issues involved. Responses to problems that emerge are then formulated in "catch-up" mode, leading people to believe, erroneously, that science and technological innovation necessarily happen first, and that social and ethical and political interventions follow behind, cleaning up.[11] This view of innovation is so pervasive that the human life sciences and biomedical sciences that have ethics have incorporated it into the dominant expression of the ethics that must necessarily accompany them in the widely used acronym ELSI, which stands for "ethical, legal, and social implications."[12] Descriptions of the acronym's early use in conjunction with the Human Genome Project emphasize the departure from an older model:

ELSI provides a new approach to scientific research by identifying, analyzing and addressing the ethical, legal and social implications of human genetics research at the same time that the basic science is being studied.[13]

It was indeed novel that people felt that ELSI concerns with human genomics should not be left until after the science had been completed but instead should be attended to, and funded, during the scientific research process. Nonetheless, in important ways the old temporality was left intact: ethical, social, and legal concomitants of research were still couched as "implications," downstream from the science itself. Stem cell research takes this "upstreaming" of the ethics of science one step further. No one considers the procurement of embryos for research to be an implication of the research; it is, rather, a precondition of some kinds of research. In the ELSI formulation of the Human Genome Project, the word "implications," separated the science from its consequences, retaining for scientists and industry a certain freedom to carry out the science in question without interference. While it is understandable that scientists—and administrators and investors—would desire this freedom for scientific and economic progress, and often deem it important for the impartiality of their work,[14] this separation between ethics and science for fields that have ethics has become increasingly untenable.[15] At the end of the beginning of human pluripotent stem cell research, both the clean-up / catch-up model of ethics lagging behind science, and the simultaneous ethics and science ELSI model are insufficient. Good science for fields that "have ethics" requires a deeper intertwining of science and ethics, and will thus require a renegotiation of what is to constitute research freedom in both.

THE PRO-CURIAL FRAME, ITS BIOPOLITICS, AND ITS NECROPOLITICS

How, then, does science and technology proceed when it is beset by ethical problems not only as downstream implications? In the case of human pluripotent stem cell research in the United States, it was enabled to proceed by the construction of what I call the "pro-curial" frame or landscape. This frame was most clearly articulated in California in conjunction with the passage of Proposition 71 and the resulting establishment of the California Institute for Regenerative Medicine. The pro-curial frame had three major elements:

operationalizing ethical objection to hESC research as a problem with the *procurement* of research material (human embryos and gametes and hESC lines), and framing solutions as a process of delineating acceptable new means of procurement

developing *curatorial* protocols and practices to track, assign custody, and certify the provenance and process of procurement

a *pro-cures* rhetoric driving innovation and investment and countering ethical dissent with an alternative ethical world view.

PROCUREMENT

Each of the major players engaged in contesting and promoting the public funding of hESC research in the United States in the first decade of the twenty-first century (including the Bush administration) operationalized ethical opposition as a problem of procurement. (See table 2.1.) Each group, however, put different conditions on how research materials—and *what* materials—could be obtained acceptably.

As can be seen from the table, the Bush administration, in power from 2001 to 2008, had a threefold strategy of procurement. (See the "Republicans" cells in table 2.1, where boldface signifies the three components that made the Bush policy distinctive.) The first part of the Bush strategy was to make the issue about what the U.S. taxpayer could be asked to fund, rather than about what research could be done. The second part of the strategy tightly controlled what research efforts could receive federal funds. After the announcement of the Bush policy in August 2001, the government's science funding agencies could fund research on hESC lines that already existed (the so-called presidential lines), but could not fund research that would require the creation of new lines. To fund research on newly created lines would imply government endorsement of the destruction of new embryos, whether or not the government funded the destruction itself. It was deemed acceptable to do research on stem cell lines that already existed because no new harm was being done, and this enabled a compromise; in general, as long as it didn't contradict the right to life, government funds should support science, especially science that promised to improve many people's lives in times to come. By permitting funding on already existing lines, this policy went further than many pro-life constituencies would have liked. It did not go far enough, however, for most supporters of the research. The third and final part of the Bush policy involved promoting research on adult stem cells. Embryos procured from fertility facilities after the date of the policy could not be used in federally funded research; the fate of "spare" embryos in clinics became a symbolic battleground. The Bush procurement policy became the restrictive background for research against which stem cell advocates pushed.

The authors and backers of California's Proposition 71, the Stem Cell Research and Cures Act of 2004, articulated a different strategy of procurement. Proposition 71 was expertly crafted to permit the funding, through new California taxpayer-funded bonds that its passage would authorize, of

TABLE 2.1

Support for public funding of human pluripotent stem cell research by procurement method or source, United States, 2001–2011.

	Republicans	Democrats	CIRM
Support use of existing hESC "presidential lines"	Yes	Yes	Yes, but lower priority than making new lines
Support derivation of new lines from "spare" embryos, IVF clinics	No—these should be surrogate-adopted	Yes	Yes
Support derivation of new lines from somatic cell nuclear transfer (SCNT) with eggs	No	No	Yes, as long as no reproductive cloning; interest decreased after Korean scandal, and iPS
Support "adult" (may be sourced from aborted fetuses) stem cell research	Yes	Yes	Yes, but low priority; Proposition 71 intent to increase funding for hESC
Support induced pluripotent stem cell research, commercial cell sources	Yes, but not very engaged	Yes	Yes, but lower priority at first, and then higher priority by 2011
Support research with non-viable embryos or biopsied blastomeres from IVF clinics	No	Not engaged	Yes, but pursued by private researchers to get around bottlenecks of supply
Support research with altered nuclear transfer (ANT)	No— proposed by Catholic PCB member	Not engaged	No
Support research with women's eggs donated for research	Not engaged; a matter of private medical ethics	Advisory, e.g. IOM / NRC 2007 report, egg-donation risks	Yes, as long as donors not paid; "Ortiz Bill" (SB 1260, 2006) added other protections; AB-926 to lift payment prohibition vetoed
Support research on animals' embryos, gametes, bodies, cells	Not engaged, but taken for granted	Not engaged, but taken for granted	Not engaged, but taken for granted

research involving the derivation of new embryonic stem cell lines. It avoided the words "embryo" and "cloning" and rhetorically melded being pro-cures with its solution to procurement. Proposition 71, and thus the resulting California Institute for Regenerative Medicine (CIRM), made a distinction between human embryos positioned within a reproductive context and those that had already exited the reproductive realm. This strategy targeted the so-called spare, supernumerary, or leftover embryos that had grabbed the headlines since the mid 1990s as *in vitro* fertilization clinics and their patients around the world pondered the fates of growing numbers of frozen embryos stored but no longer needed for infertility treatment.[16] Stem cell research needed human embryos, and fertility clinics had too many; moreover, many of these leftover embryos were going to be destroyed or donated to research anyway, and donating them to stem cell research could save lives down the line. Not only were certain leftover embryos (which could be obtained with the consent of the patients who had custody of them) available in problematic abundance, and already certified as not destined to lead to the birth of a child; their use also gave the promoters of stem cell research their own pro-life rhetoric. The work of taking an embryo out of a reproductive trajectory had already been done and so would not have to be done by stem cell researchers themselves, and researchers would turn this gift of embryos into life-saving therapies.

Proposition 71 did not mention embryos, because its backers wanted the public to vote for it. The conceptual rationale for this was that the zygotes in question were no longer in a reproductive context and so were already not functioning as potential human beings in the sense of being on a path toward personhood. Backers also avoided the word "cloning" for the purposes of public support for the proposition, except for the explicitly banned "reproductive cloning," from which they differentiated the research cloning or somatic cell nuclear transfer (SCNT) procedure that could receive funds.[17] It was not until after the establishment of CIRM that the question of ethical protections for the egg donors who would be needed for somatic cell nuclear transfer research came to the fore.[18] This contributed a piece to the pro-curial landscape, though, by establishing a fundamental difference between those who donated bodily materials for stem cell research and those who funded and engaged in the research in terms of who could profit from the state stem cell initiative, deeming only the latter groups part of the economic value chain. Women could not be paid for their eggs but stem cell research was going to make money for California.[19]

The Democratic position after the election of Barack Obama in November 2008 followed attempts by Democratic politicians during the Bush administration to change the 2001 policy on the federal funding of stem cell research. Between 2004 and 2008, two different bills made their way through Congress, and George W. Bush exercised his first presidential veto on one of them. The Democratic position on procurement was not as permissive as the CIRM one, taking on the same argument for being able to use embryos left over from completed fertility treatment, but without promoting the research through any constitutional amendment, without earmarking special funds, and without proposing to permit somatic cell nuclear transfer cloning. Spare embryos became the focus of national political theater during this period. In remarks made on May 24, 2005, President Bush recapitulated his 2001 policy, and brought in as star witnesses a number of "snowflake" children, born after being adopted as embryos and brought to term through gestational surrogacy by their adoptive mothers:

[I]n August 2001, I set forward a policy to advance stem cell research in a responsible way by funding research on stem cell lines derived from embryos that had already been destroyed. This policy set a clear standard: We should not use public money to support the further destruction of human life. . . . The children here today remind us that there is no such thing as a spare embryo.[20]

This speech was delivered on the day that the House of Representatives passed the Stem Cell Research Enhancement Act of 2005 (H.R. 810), a bill sponsored by Representative Michael Castle (D-Delaware), by a vote of 238 to 194. The bill then was passed by the Senate (as S. 471, sponsored by Senator Arlen Specter, R-Pennsylvania) by a vote of 63 to 37 on July 18, 2006. President Bush vetoed the bill, with his first veto, on July 19. The House of Representatives was unable to override the veto, which would have required a two-thirds majority (the vote, on July 19, 2006, was 235 to 193). A similar fate awaited the second such bill, the Stem Cell Research Enhancement Act of 2007 (H.R. 3, sponsored by Representative Diana DeGette, D-Colorado; S. 5, sponsored by Senator Harry Reid, D-Nevada). Despite enjoying a large majority that presumably represented widespread support for hESC research in the United States, these Democrat-sponsored bills did not become law. They did, however, set the stage for the Democratic position that was ready and waiting after the election of Barack Obama. The bills established a different place to draw the line between acceptable and unacceptable, permitting the use of leftover embryos for pluripotent stem cell research. In the eyes of many, the bills also provided a pedigree of support for and belief in science relative to the Republicans. Obama did not articulate a new policy, but instead removed the Bush policy, allowing clinics that offered couples the choice to donate leftover embryos to research to do so without prejudicing funding to the subsequent research. This was, in some sense the opposite of Proposition 71, which was aggressively activist. The Obama administration's position allowed more of the issues around procurement to slide into the background, covered by general federal and local codes for informed consent and for the procurement of human tissue. The Obama administration's repeal, however, was accompanied by a new rhetoric of affirming belief in science, and by a reformulation of the President's Council on Bioethics to the Presidential Commission on Bioethical Issues that would begin to reflect the deliberative democratic ethics of mainstream American bioethics.

During this time period, others proposed different solutions to problems of procurement of human pluripotent stem cell lines. A few researchers suggested using more readily attainable "non-viable" embryos to overcome the scarcity of embryos available for research.[21] Robert Lanza's group at Advanced Cell Technology used a procedure similar to pre-implantation genetic diagnosis to remove a single blastomere without destroying the whole blastocyst, and then derived stem cell lines from the blastomeres.[22] I advocated, from a women's-health perspective, using embryos especially made for research, or embryos from somatic cell nuclear transfer, rather than spare embryos, so as to remove the influence of stem cell research on the conduct of fertility treatments and pre-implantation genetic diagnosis.[23] A Catholic member of President Bush's Council on Bioethics suggested a process he named "altered nuclear transfer": altering the gametes before fertilization so that a resulting embryo would not be viable.[24]

The Catholic Church distinguished itself from the Evangelical Protestants in the US by actively promoting—as opposed to advocating it over embryonic stem cell research—various kinds of adult stem cell research, a pattern that has continued.[25] Specificities of the Republican position during the stem cell debate over this time throw light on the extent to which it reflected a Protestant Evangelical version of pro-life views, rather

than a Catholic position that was a more common source of ethical opposition to stem cell research in other countries around the world with
restrictive stem cell policies.[26] Not only have Catholics advocated more
strongly for adult stem cell research; many, taking their lead from the
Vatican, did not condone the adoption of "snowflake" babies, and took a
much harder line than most right-wing Evangelical Protestants in the
United States on assisted reproductive technologies in general.[27]

Table 2.1 illustrated the front lines of dissent, but it also highlighted some
of the paradoxes that have arisen in how procurement was negotiated. Pro-
life groups advocated adult stem cell research, despite the fact that often
"adult" (i.e., non-pluripotent) stem cells are taken from aborted fetuses, and
despite the fact that this tissue is regularly purchased privately. A trend from
2008 to 2011 to exempt most research on induced pluripotent stem (iPS)
cells from stem cell research oversight, because no embryos or eggs have
to be procured, moved much of the procurement ethics of stem cell research
into the older (and, I would argue, inappropriate) ethical terrain of waste,
gift, and commercial human tissue exchange.[28] Protecting egg donors originally rested on denying payment to egg donors, while setting up stem cell
research so that everyone else involved in the research could and should
seek to be enriched in economic and symbolic capital. California Assembly
Bill 926 of 2013, "Reproductive health and research," would have made it
possible to begin to pay egg donors in California more than direct costs,
but was vetoed by Governor Jerry Brown.

CURATORIAL PRACTICE

It was not enough simply to settle (enough to get on with research) the
procurement principles delineating acceptable research materials. It was
also necessary to develop the bureaucracy for attesting to this acceptability,
and, to some extent, to the cells' appropriate uses thereafter. In California,
as elsewhere, a densely *curatorial* practice, combining bureaucracy, care, and
authority over interpretation and circulation of products, arose. For research
funded by CIRM, for example, one institutional entity and two concepts
became particularly important in this curatorial practice. Every institution
receiving or hoping to receive CIRM funding for pluripotent stem cell
research, whether public or private, had to have a Stem Cell Research
Oversight Committee (SCRO). A major job of such a committee was to
oversee the "acceptable" derivation of "covered" cells. The concepts of

"acceptable" and "covered" were reactions to, and work-arounds of, anti-abortion concern. Together they show how ethical contest was operationalized into curatorial practice.

Under CIRM Medical and Ethical Standards regulations, the relevant information is to be found in the following sections of the California Code of Regulations, Title 17:100070 (SCRO Committee Review and Notification), 100080 (Acceptable Research Materials), and 100090 (Special Considerations for CIRM-funded Derivation).[29] These sections have been revised frequently, with public comment periods, and their revisions reflect evolving science and evolving ethical practice and legal argument on the issues at hand. By the amendment of February 2010, SCROs had to undertake a full review of protocols that proposed to "procure, create, or use human gametes or embryos." They also had to undertake a full review of proposals that included transplantation of "covered" cells to non-human animals. This iteration of the regulations emphasized its attempt to bring CIRM in line with National Academies of Science Guidelines. In particular, everyone was working to remove the use of somatic cells for iPS experiments from high levels of scrutiny.[30] Now iPS research required SCRO notification only if using identifiable cells, and the lowest level of oversight (a statement of compliance) if using de-identified cells, even if they came from fetal tissue, as specified in California Code of Regulations, Title 17, Section 100085.[31] Human and animal subjects consent need only be in line with that applicable to all kinds of research.

The concept "acceptable" was curatorial from the start, and did not mean acceptable to anyone in particular, let alone to everyone. For CIRM, it came to mean that a particular stem cell line was procured from a source or created using a method that could be interpreted as having been approved in the California Stem Cell Research and Cures Act. Embryonic stem cell lines could be acquired from any CIRM "authorized authority," as long as any embryos or gametes used to make the line were properly consented, and as long as any embryos used were leftover from fertility treatment or the result of SCNT. Over time, stem cell lines from different bio-banking facilities around the country and the world, as well as the human embryonic stem cell lines on the National Institutes of Health hESC registry, were included among those lines considered from CIRM's perspective to be acceptably derived. The word curator ranges in its meanings from country to country, but it includes selection and care of, as well

as guardianship, transport, exchange, and interpretation of, culturally valuable things whose value cannot be reduced to their price on any market.
The concepts of acceptability and being covered perform this work for
hESC research.

PRO-CURES ASPIRATION

A search for cures as a rationale for the public funding of science has a
long history at the federal level in the United States; one example is the
"war on cancer" launched in the 1970s. Promoters of human embryonic
stem cell research also developed a pro-cures rhetoric, but one that reached
its apogee not in the demand for federal funding but in California's Proposition 71, the California Research and Cures Act. This pro-cures rhetoric[32]
cast supporters of stem cell research as fighting to cure pervasive and
devastating diseases and medical conditions. By equating the fundamental
ethical imperative to save and improve lives with the provision of state
funding for embryonic stem cell research, the pro-cures rhetoric counter-
balanced pro-life objections to the research:

> The people of California find and declare the following: Millions of children and
> adults suffer from devastating diseases or injuries that are currently incurable,
> including cancer, diabetes, heart disease, Alzheimer's, Parkinson's, spinal cord inju-
> ries, blindness, Lou Gehrig's disease, HIV/AIDS, mental health disorders, multiple
> sclerosis, Huntington's disease, and more than 70 other diseases and injuries.
> Recently medical science has discovered a new way to attack chronic diseases and
> injuries. The cure and treatment of these diseases can potentially be accomplished
> through the use of new regenerative medical therapies including a special type of
> human cells, called stem cells. These life-saving medical breakthroughs can only
> happen if adequate funding is made available to advance stem cell research, develop
> therapies, and conduct clinical trials.[33]

To make plausible this ethical claim that the point of the research was
cures, the research had to be shown to be concerned with the entire
innovation trajectory, all the way from as-yet-undone basic science to
clinically valid treatments. The common use of the expression "bench to
bedside" in connection with stem cell research, and the emphasis on
"translational" research, did just this. The bench-to-bedside commitment
also lent itself to being read as a commitment to funding a new field of
innovation, putting California out ahead of the rest of the U.S. and even
the world. State investment in this "next Silicon Valley" had the potential

to reinvigorate California's economy; the research might also dramatically cut medical costs currently incurred by dealing with chronic conditions that might be cured with stem cell research. Innovation and cures were bundled together as promise and potential, and potential was what these pluripotent cells had in abundance.

The beginning of human embryonic stem cell research in the United States faced actual or potential market failures that the pro-curial landscape of being pro-cures, solving the procurement issues, and developing curatorial practices had the potential to solve.[34] The first market failure related to the source of the original ethical controversy: the objections of pro-life constituencies to using federal funds to support research that needed human embryos as raw materials and that then destroyed those embryos in the process of extracting stem cells. These objections had been national policy since President Bush's 2001 policy. Denying federal funding for this field of scientific research split the polity in two in two ways. First, it separated the private sector (which was not affected by the ban on funding) from the federal public sector as a source of funding for basic science research. Second, it separated the individual states from the nation as a source of funding for the research. This pushed debate about funding and setting up of ethical and regulatory equipment to jurisdictions that were not accustomed to being science policy leaders, namely, individual states and private capital. Both states and private capital had long been players in supporting and funding research, especially biomedical research, but they had not typically done this in a vacuum of federal funding (whether from military or civilian research funding bodies), especially for the basic science part of the innovation chain, and they had not typically acted in opposition to federal policy, being more used to a federal lobbying role. At the same time, the public lost many of the assurances, imperfect as they might be, that come with federal funding that scientific excellence and ethical conduct would be promoted.

In the private sector, other market failures were evident, although several of them were further down the innovation pipeline than procurement of the original stem cells themselves.[35] The potential of human embryonic stem cell research seemed clear to its pioneers who foresaw that these "cell lines should be useful in human developmental biology, drug discovery, and transplantation medicine."[36] Using stem cells to improve our basic understanding of developmental biology seemed relatively straightforward conceptually if not in practice: embryonic stem cell lines provide an immortalized, thus regenerating, supply of undifferentiated cells upon which to experiment to uncover the mechanisms of cellular differentiation. But using stem cells for drug discovery or for transplantation medicine (for cures, the bedside half of the bench-to-bedside promise) held many challenges. For drug discovery, the cells held all sorts of potential. They offered the possibility of doing a lot of testing that is currently done in animals or in human subjects in a dish on the appropriate target cells.[37] The potential availability of embryos that had been deselected during pre-implantation genetic diagnosis because they carried a gene linked to a disease suggested that stem cells from those embryos could be miniature models of the disease in question, or, as it became known, a "disease in a dish," on which toxicity and efficacy drug discovery testing could be carried out.[38] Despite this great potential, it would be necessary to find ways to standardize the cells sufficiently to produce reliable results, and ways to bring drugs to market that had come through a discovery and testing process in a dish rather than through the usual route (*in vivo* animal studies and human testing). For example, if the whole point was to change the process of bringing drugs to market in the first place, would it still be necessary to carry out the required animal testing for an IND (investigational new drug) application to the Food and Drug Administration so as to proceed to clinical trials?

Transplant medicine also held great potential from the start but had its own problems. One was the specters of gene therapy and of cloning, which would hold back public support, and which also suggested that efficacy and safety hurdles would be formidable.[39] Another challenge was the nature of the therapeutic substance itself. Unlike drugs made up of small molecules, human biologicals are hard to standardize, and fit poorly with the current intellectual-property regime in the United States.[40] One of the most exciting things about the potential of embryonic stem cells for transplant medicine is the potential to make transplantable cellular therapies starting with cells from a patient's own stem cells, autologous transplants. A patient's own cells are histocompatible and so would reduce the problems presented by tissue rejection and graft versus host disease in transplant medicine.[41] Which parts of the development and delivery of customized transplants could be scaled up and standardized sufficiently to be of interest to the private sector, however? In short, new business models and new paths for bringing pluripotent stem cell products to market would be needed, especially if being pro-cures meant being pro-innovation, and vice versa.[42]

For all these reasons and more, U.S. venture capital did not rush to fund embryonic stem cell research. It seemed unlikely that returns would be forthcoming in a reasonable investment time frame, and the pharmaceutical industry, which had defined and dominated both the testing and business models of biomedical R&D that were already in place, did not simply engulf the field because of its poor fit with the preexisting R&D and intellectual-property apparatus.[43] Indeed, the corporate part of the private sector to some extent saw itself in need of its own subsidies to assume the amount of risk the bench-to-bedside research trajectory of embryonic stem cell research would involve, especially given the withdrawal of federal public funds from the early stages of research, and for-profit institutions were eligible to receive CIRM grants from the beginning. Private funding— the kind not affected by Bush's restrictions on federal funding—also encompassed funding channeled from patient advocacy organizations and from wealthy philanthropists. These two sources of funds (patient advocacy organizations and philanthropists) were extremely important to the pro-cures rhetoric of Proposition 71 and CIRM. From the crafting of Proposition 71 to the latest World Stem Cell Awareness Day, patient advocates have been the public face of California's stem cell initiative. Philanthropists put up the money for getting Proposition 71 on the ballot and tided it through until the legal challenges to its existence had been resolved and the state bonds became available. Market failures, the problem facing for-profit investors in the research, were the problem these two groups were in part fixing. That did not mean, however, that these two groups were ideal funders for the research in the absence of major sources of public funding. They both came with their own shortcomings as funders of pluripotent stem cell research.[44]

"Billanthropy" (a kind of philanthropy named for the Bill and Melinda Gates Foundation's unprecedented wealth and the scale of its influence in various kinds of research funding including the biomedical) is able to redistribute excess capital accumulation to targeted projects without a public mandate and with little oversight as to the projects supported so objections to stem cell research did not stand in the way. Not only is philanthropy able to target its donations; it is tax exempt. Supporting stem cell research through charitable giving thus diverts from federal and state coffers what would otherwise be substantial tax revenue on that money, and gives privately wealthy individuals and their foundations the power to spend on projects such as the search for cures that their tax dollars could not support. Some see philanthropic giving as an essential way around the flaws of a two-party democracy. On the other hand, one could argue that controversial yet highly promising fields such as stem cell research, even more than less controversial fields, should benefit from public funding and its concomitant public oversight. With a paucity of federal stem cell funding, and with the health-care system in crisis, philanthropic and other private donations stepped in to fund not just the research itself and positive messaging campaigns for the stem cell initiative, but all parties to the vibrant stem cell debate from progressive voices lamenting the threat posed by stem cell research to our human futures to socially conservative abortion foes (though to very different levels; see table 2.2).

Multiple voices are part of democratic debate, so it is hard to "just say no" to private donors. Furthermore, given the rising costs of research, and against a backdrop of the normalization of ever-closer ties between public research institutions and the private sector, private donations can seem to be an essential way to maintain research quality and competitiveness. Entire apparatuses of enticement to wealthy private donors are as characteristic of modern U.S. research universities as their less wined and dined sponsored projects offices are for channeling federal government funding to campus. In the context of California's economic slowdown and the end of the housing bubble, and the shockingly inequitable Proposition 13,[45] many research institutions feel they have no choice but to court private donors. By the time of the passage of Proposition 71, California's public university system had already begun to require fundraising capability, and not just academic excellence and educational vision, of its senior management, and many of us (myself included) participated in private fundraising for campus stem cell research efforts. For all kinds of reasons, then, philanthropic donors and foundations, though problematic as science funders, became standard bearers for stem cell research in California.

If the logic of innovation was retained through these substitutions of philanthropy and the state for the federal government, what role was assigned to the California taxpayers, now footing the bill for state-supported research? The replacement of federal money with state funding and private donations should have been mirrored by a replacement of the

TABLE 2.2

Data on Proposition 71, the 2004 ballot question "Should the 'California Institute for Regenerative Medicine' be established to regulate and fund stem cell research with the constitutional right to conduct such research and with an oversight committee? Prohibits funding of human reproductive cloning research." Passage of this proposition amended the California Constitution by adding Article XXXV. Sources: Official Voter Guide to Proposition 71; Analysis of Fiscal Impact by the Legislative Analyst's Office; Summary by Attorney General; November 2004 Election Results (California Secretary of State); archived websites of "Yes on 71" and "No on 71" (available at http://www.ballotpedia.org).

	Major supporters	Other data	Votes
Yes	Coalition for Stem Cell Research and Cures, which included 22 Nobel laureates; celebrity patient advocates Christopher Reeve and Michael J. Fox; more than 50 patient advocacy organizations; many other business and academic and medical signatories	Total raised: ~$34 million Major donors: Robert Klein ($3 million) John Doerr ($1.9 million) Marion and Herb Sandler ($1.18 million) Juvenile Diabetes Research Fund ($1 million) Pierre and Pamela Omidyar ($1 million) Gordon Gund ($1 million) William Bowes ($600,000) Bill Gates ($400,000)	7,018,059 (59.1%)
No	Roman Catholic Church Orange County Republicans California Pro-Life Council Pro-Choice Alliance Against Proposition 71 Taxpayers for Fiscal Responsibility	Total raised: ~$400,000 Major donors: United States Conference of Catholic Bishops ($50,000) Howard Ahmanson Jr. ($95,000)	4,867,090 (40.9%)

general national public who pay federal taxes with the patient advocates behind private donations, but also with the general public of the state in question. In California, patient advocates became proxies for the public, and the state's taxpayers struggled to find venues in which to be represented. Patient advocates, as many scholars have shown, have to be both professionally organized and expert.[46] They also interact in importantly symbiotic ways with the research establishment to attract funding and focus research upon the diseases for which they advocate. As a matter of the greatest urgency, often literally a matter of life and death, patient advocates mobilize a kind of neoliberal subjectivity, or "biosociality."[47] With its self-governing ethos, patient advocacy does not intend to speak for everyone, and absence of a market rationale for research into a cure (e.g., for "orphan" diseases) or a lack of federal support (as in stem cell research) makes the need for private funding even greater. In the field of stem cell research, patient advocacy organizations are frequently noticeably less ambivalent about prospects for cures than scientists themselves, for example.[48] In California, Robert Klein, a wealthy real-estate developer whose teenage son had diabetes, crafted Proposition 71 and wrote into it the structure of representation and oversight that its passage would put in place. The proposition was then promoted by the "Yes on 71" campaign, a coalition of patient advocacy organizations and medical organizations that outspent the various groups opposing Proposition 71 by approximately 85 to 1. The California Legislative Analyst's Office estimated that it would cost $6 billion over 30 years for the state to pay off the principal and interest on the $3 billion worth of general obligation bonds to be paid for from California's General Fund.[49] Although the proposition was the result of "direct democracy" (a citizen-led proposition), governance of, by, and for the people of California faded.

In a strange elision, patient advocates' interest in funding the research came to be seen as qualifying patient advocates, rather than disqualifying them, for the task of providing "lay" or "public" input to the new infrastructures of stem cell funding and research that were being put in place. As stipulated in the proposition, the leadership of the state agency to be founded, the California Institute for Regenerative Medicine (CIRM), was also to be undertaken by interested parties rather than more neutral parties. CIRM's governing body, the Independent Citizen's Oversight Committee (ICOC), was required by the constitutional amendment to be composed of representatives of leading research institutes, biotech companies, and patient advocates, rendering it neither an "independent" nor a "citizen's" oversight committee. The presence of leaders of research institutions both public and private—and thus likely recipients of CIRM funding—on the ICOC became a focus for concern about potential conflicts of interest in the agency,[50] but patient advocates avoided similar scrutiny, perhaps because their motives are to seek cures rather than to attract grant money or make money *per se*.[51] However untainted by financial gain and pressing the motives of patient advocates may be, their interests are to fund research that leads to cures in time to help, not to fund research equitably, or any other such qualification to their basic remit.[52] Advocacy and stakeholdership became proxies for the public good in the absence of federal funding; in turn, state funding was possible because national policy was seriously at odds with the views of the majority of the general public in California.

BIOPOLITICS AND NECROPOLITICS OF THE PRO-CURIAL FRAME

Human pluripotent stem cell research, then, was launched in the United States as a science that "had ethics." In a climate of serious restrictions of federal funds for research, a combination of private and state funding converged around what I described above as a "pro-curial" frame or landscape characterized by principled, if contested, procurement parameters, curatorial practices for accounting for acceptably derived research materials and cell lines, and a pro-cures rhetoric of innovation. Implicit in this innovation landscape, as in any other, are both biopolitics and necropolitics. It is important ethical work to draw out these bioscapes and necroscapes, in stem cell research. The risk is that ethical attention might fade with the normalization of research, rather than being integrated into research, ever demanding anew the incorporation of more people and more issues as the research progresses.

The term "biopolitics" is deployed in many ways.[53] In one usage, it refers to the politics of things "bio," encompassing social and political debates having to do with any aspect of biotechnology, biomedicine, or bioethics.[54] This use of the term alerts us to the fact that the life sciences have politics, and that sometimes certain people and countries and institutions gain (sometimes a great amount) and some lose (sometimes everything) in these politics. In another usage, loosely following Foucault, biopolitics is the politics involved in exercising "biopower,"[55] the term used for the rational, non-sovereign form of power exercised by a modern nation state when it exerts control over populations through managing such things as public health, reproduction, eugenics, and race.[56] In other work, "biopolitics" describes who is allowed or selected to live (or to live well or rule), and who to die (or be sacrificed or vivisected or not counted) through the management of the life of populations. The concentration camp and National Socialism, the plantation and slavery, the West Bank and Middle East politics, the colony and the metropole, and AIDS and compulsory heterosexuality, for example, have been examined as the scenes of biopolitical abjection (or necropolitical normal) that is in some way definitive of modern state power.[57] Each of these different approaches to biopolitics ask, in one way or another, who lives at whose expense through which technics, and how death's attendant violence is normalized and its victims dehumanized; who can be mourned, who has a biography, and the taking of whose life or quality of life is a crime?[58]

The pro-curial landscape for conducting pluripotent stem cell research can be thought of as a biopolitical landscape: to conduct pluripotent stem cell research in California was to imagine, and then work toward, saving and/or ameliorating countless lives. It was also to imagine regenerating the economic life of the state through getting out ahead in this sector, and to re-commit to the epistemology and fruits of science. The Yes vote on Proposition 71 made funding this pro-cures-as-innovation vision a matter of direct government action. Many in biotech and related businesses, as well as many patients, politicians, scientists, doctors, democrats, and their allies, were made lively by this biopolitics.[59] From the point of view of those who opposed the research on the grounds that it destroyed human embryos, stem cell research offered a necroscape, dealing in death worlds of the unborn. Once the pro-cures-as-innovation package had been promulgated, it cast pro-life motivated restrictions on federal funding as its own kind of necroscape, casting those suffering from potentially curable conditions as the "living dead" condemned to "living in prognosis," diagnosed by but not treatable by the biomedical and bio-informatic terms for understanding contemporary bodies.[60]

California's pro-cures landscape, as much as President Bush's "snowflake" one, explicitly operated with a mandate to save lives, each the other's necropolitical counterpart, pitting cures against life of the unborn. The pro-cures landscape used principled procurement and curatorial practices (getting cells only from particular provenances, and handling them according to particular guidelines).[61] The "pro-cures" and "snowflake" bioscapes were defined in relation to one another. This pairing, however, masks other possible stem cell bioscapes that do not center directly on protecting or overcoming barriers to using human embryos, and which, for this very

reason, take more effort to bring to light. Four major such alternative
bioscapes struggled for articulation as I carried out research during the first
phase of human embryonic stem cell research in the United States, especially California:

the relations between the funding of human pluripotent stem cell research
and disparities in social justice and health
tensions between the emphasis on cures and other aspects of disability
justice
a paradigm shift occurring regarding research subjects and body part
donors arising from the progress of fields such as stem cell research
the "dual use" potential of stem cell research to become a destructive and/
or defensive, rather than a regenerative and/or prospective technology.

I turn first to the question of the U.S. debate over human embryonic
stem cell research, and its relation to broader questions of social justice
and health disparities mitigation. To think about this question is to ask
about hESC research's external biopolitical others in the United States
and California—those whose lives are potentially impacted despite being
outside of the pro-curial landscape and so not enlivened by its success.[62]
Had Proposition 71 not passed, the monies now spent on stem cell
research in the state could have been spent in other ways, perhaps even
helping to lessen the state's huge budget shortfall and to keep public
research institutions running.[63] The pro-cures vote presented a choice
between supporting the science and not supporting it, but it was also
about the state's funding priorities. Many hoped that California's support
for hESC would benefit the state. Three kinds of possible fiscal pay off
were widely debated by us all during the run-up to the election: that
there would be economic activity as a result of the investment that would
provide jobs and increase tax revenues (a stimulus package argument with
subsequent trickle down); that it would result in cures and treatments for
currently incurable diseases and decrease the costs to the health-care
system of chronic disease and disability; and that there might be patent
activity and royalty streams, some of which could be earmarked for the
state. Many thought that cost saving was unlikely (what new medical
treatments have ever brought the costs of health care down?) and that the
other possibilities had long time horizons.[64] It was also unclear how avail-

able any CIRM-funded treatments would be to the uninsured, and thus unclear whether the promised cures—if they emerged at all—would be available only to people with adequate health insurance. The people of California, as well as the drafters of Proposition 71, voted to fund embryonic stem cell research but not access to health care.

With pro-life litigation against CIRM ongoing,[65] and as part of my efforts in our nascent Berkeley Stem Cell Center to take advantage of state-wide resources and attention on stem cell research to introduce other kinds of ethical issues to the mix, I met with Dr. Zach Hall, president of CIRM, to discuss our upcoming Toward Fair Cures conference.[66] We had conceptualized the conference around the question of the obligations to a diverse public that should flow from a public vote for and taxpayer funding of a state bond issue to finance stem cell research. We, the conference's organizers, were arguing for such things as scientific workforce diversity, for an equitable sharing of state-funded stem cell research earnings (should there be any), for critical attention to the use of ethno-racial categories in stem cell research and the use of disabilities in research promotion, and for the development of plans for access to therapies that resulted from California taxpayer funds. Addressing these issues of economic and social justice did not help with the embryo-destruction issue, however; neither did it necessarily advance innovation in the field, which many saw as the main way in which non-patients as well as patients would gain from California's head start on stem cell research. My discussion with Hall concerned whether or not the issue of justice in California's stem cell research endeavor was too far afield from CIRM's mandate. Since the voters had approved a program to finance human embryonic stem cell research that had difficulty being funded elsewhere, and that was aimed at therapies, money should not be spent unless such expenditure demonstrably advanced the progress of the (ethical) search for cures from bench to bedside. Furthermore, given that these justice issues have not been solved for U.S. health care in general, which remains highly stratified, it should not be required of stem cell research to solve them alone. CIRM, as Hall put it in our conversation, could "not reform all of health care." Over the next few years, the arguments in favor and against introducing social justice goals were debated, though they were very quickly operationalized as affordability and returns to the state rather than being expressed in more foundational terms. Many of us also argued for grants to educational and research institutions where diversity was a goal.[67] Americans For Cures[68] continued to be the voice of the pro-cures bioscape, expressing its rejection of the portion of 2008's California Senate Bill 1565[69] that was targeted at making therapies affordable:

While we passionately support the goal of healthcare that is accessible and affordable to all Californians, we oppose Senate Bill 1565 (SB1565). . . . SB 1565 will keep entrepreneurs, private investors and big companies from developing CIRM-based discoveries.[70]

Much progress was made, culminating in the passage of Senate Bill 1064 in the fall of 2010.[71] SB 1064 included the following approaches to affordability, earnings to the state, and commitments to in-state purchasing:

It is in the best interests of the state that therapies that are created in whole or in part by funding from the institute be made available to Californians who have no other means of purchasing those therapies for reasons that include, but are not limited to, low income or the lack of available health insurance coverage. . . . The ICOC shall establish standards that require all grants and loan awards be subject to intellectual property agreements that balance the opportunity of the State of California to benefit from the patents, royalties, and licenses that result from basic research, therapy development, and clinical trials with the need to ensure that essential medical research is not unreasonably hindered by the intellectual property agreements. . . . The ICOC shall establish standards to ensure that grantees purchase goods and services from California suppliers to the extent reasonably possible, in a good faith effort to achieve a goal of more than 50 percent of such purchases from California suppliers. . . . A requirement that each grantee or the exclusive licensee of the grantee submit a plan to CIRM to afford access to any drug that is, in whole or in part, the result of research funded by CIRM to Californians who have no other means to purchase the drug.[72]

Though truly significant, the bill modified but did not disrupt the pro-curial landscape, negotiating returns to the state according to intellectual-property agreements premised on future earnings from successful innovation. It would have been a much more ambitious task to promulgate an entire bioscape aimed at funding hESC for economic and social justice and the mitigation of disparities in health.[73]

I turn next to a second possible alternative bioscape, that of disability justice. The pro-curial landscape is centrally concerned with disability. It serves to garner resources for, and aligns institutions, rhetoric, and actors toward, finding biomedical solutions to debilitating illnesses and disabilities. Many people with injuries and disabilities have committed enormous amounts of time, money, and effort toward making stem cell research legal and toward speeding up the translational process from basic science to safe and effective therapies. When the process is slowed down, as when President Bush restricted access to federal funds for hESC in 2001, it can be taken as a betrayal of patients who might have been helped (sooner) by the fruits of research. When the Geron Corporation withdrew from the first partially CIRM-funded hESC FDA-approved clinical trial in the fall of 2011, before the trial was complete, it was seen by some as a betrayal. Although Geron refunded CIRM, this particular betrayal had a distinctly ethical dimension because of the pro-cures mandate: CIRM's funds were loaned to Geron for the purposes of advancing therapeutic applications, and by ending a trial before enough people had been enrolled to make its results statistically meaningful, for putatively purely financial motives, the company unbundled the search for cures from the biomedical innovation process. Arguably, the trial should not have been allowed to fail.[74] The abrupt ending for financial reasons of Geron's clinical trial reminded the public that although Proposition 71 had succeeded in presenting a bioscape where financial and biomedical regeneration were tied together, the two did not necessarily walk in lockstep.

At the same time, the pro-cures perspective scripts disability as a predominantly medical phenomenon, even though disability is embedded in hierarchical social landscapes that include pre-natal and peri-natal environments, nutritional status, pathogen exposure, lifestyle, classifications of normality, toxin exposure, and accident risk. In other words, the pro-cures bioscape contributes to the biomedicalization of society (discussed in chapter 1), and draws on the medical model of disability.[75] Most people would agree that conditions should be tackled biomedically, even if they have complex social dimensions, when the results are likely to be effective and lasting and when many of those living with the conditions in question advocate pursuing biomedical interventions.[76] Disabilities and diseases cover a vast range of the visible and the invisible, the cognitive and the physical, the chronic and the fatal, and so on; not everything can be cured, and not everything needs to be cured. Paradoxically, making cures the self-evidently desirable goal seems to rely on an unquestioned assumption that there is a cured "normal" that everyone could or should be.[77] A reductive focus on biomedical cures risks taking attention away from disability-justice activists' calls to make life rich and fulfilling for all and to deal with the fundamental inequalities that exacerbate many aspects of disease and disability and perpetuate health disparities.[78]

In terms of the rise of pluripotent stem cell research and its place in the history of science and medicine, "pro-cures" as a rhetoric also over-promises, not just in the sense of "hype" discussed above, but also in terms of the coming biomedical revolution. In a time of personalized medicine, what would it be, for any of us, to be "cured"? There is something disingenuous about touting "cures" as the rationale for stem cell research and as the desired end point for patients turning to these technologies when regenerative medicine is a part of the current revolution in personalized medicine, in which humans are being reconstructed as bio-informatic selves who live in constant prospective statistical relation to their own pending morbidity and mortality. The whole point of personalized medicine—its justification and its market structure—is that there is no longer a simple ontological distinction between the sick and the cured, or between living with a disability or disposition or not having a disability or disposition to disease; instead we have to choose which people to bring into the world, and how best to steward our own genetic lot in life once in the world. It is hard to imagine how stem cell research in a U.S. context could be conceptualized differently, however, and how the ethical choreography could be made more complex. "California Stem Cell Research and a Biomedical Contribution to a Comprehensive Approach to Disability Justice Act" just doesn't have the same ring to it as "Stem Cell Research and Cures Act."

A third bioscape gaining momentum during this period, but in imperfect alignment with the pro-cures vista, concerned a gradual but pervasive shift in the concept of a research subject, and the appropriate relations between research donors and their donated gametes or tissue. Who is a research subject? Should research donors be compensated, and if so, how? How should research donors and subjects be protected? What should be done with their tissue and its value-added products, and what with the bio-information derived from their bodies? I take up these questions as ethnographic-philosophical inquiries in part III of the book, focusing first on the questions raised by the possible subjectivity of the embryo and the chains of custody, consent, contact, and compensation for pluripotent tissue and then on the role of the animal subject and the less-than-human subject in substitutive biomedical R&D. It was not only the notion of research subject that was being re-worked; the related concept of the research donor proved equally central to this general shift in the paradigm governing research participation. During the early phase of human embryonic stem cell research in California, as elsewhere, coming to acceptable arrangements for the procurement of women's eggs proved to be of great importance, even though it did not fit cleanly with the debate over cures versus embryo protection, and even though women were not even mentioned in Proposition 71. Before the possibility of reverse engineering to induce multipotency or pluripotency, getting the eggs required for SCNT and hESC research necessarily involved women and their bodies and their biological products. The procedure for "harvesting" eggs is difficult and carries poorly understood risks to women; echoes of the buying and selling of babies and the legacies of slavery made others reluctant to contribute to processes of commodifying human life. Feminists and other activists and politicians, myself included, worked to protect egg donors and to raise awareness of the role of gender, race, class, and nation in the rise of research participation as a form of alienable bodily labor.[79] The pro-curial frame evidenced in Proposition 71 did not engage with these shifts, but it did provide impetus for them; because of the risk of exploitation of egg donors, versus the apparent unfairness of allowing scientists and corporations to enrich themselves while requiring research donors to evidence the right morals by only participating on a gift basis, it became a site where the need to move beyond earlier research donor and research subject paradigms was urgently felt. These two bioscapes, pro-cures and research donor protections, were at odds, but interacted vigorously.

Once it became possible to obtain pluripotent stem cells by induction from somatic cells, this vigorous interaction between the bipolar embryo debate and the need for research donor protections faded somewhat into the background. Despite evidence that U.S. society would continue to grapple with the evolving roles for research participants from tissue and gamete donors to surrogates to clinical trials subjects to organ donors at home and abroad, the advent of iPS cells threatened to take pluripotent stem cell research out of the picture.[80] The intensive procurement and curatorial practices developed for embryo and egg derived stem cell lines were not required of iPS cells because they started with somatic rather than embryonic or germ-line cells; avoiding the need for such procedures, along with overcoming the shortage of human eggs and embryos, was a major motivating factor for the invent-around they represented. Then as now, scientists using iPS cells were able to get the starting cells they needed for their work through established tissue-donation protocols or by purchasing of human cell lines from for-profit companies, without opening up questions about the acceptability of this procurement.[81] CIRM and our institutional SCRO committees adjusted our review policies to accommodate iPSC research; CIRM-funded iPSC research only needed to show that informed consent had been attained for the original samples. Consider the following excerpt from a 2009 paper about CIRM-funded research:

Written approval for all somatic derivations and subsequent iPSC generation performed in this study was obtained for the . . . Institutional Review Board . . . and the . . . Stem Cell Research Oversight Committee . . . , and written informed consent was obtained from each individual participant. . . . A somatic bank of nine primary adult dermal human fibroblast (HUF) lines were derived and used in this study. The gender, age and disease status/phenotype of the participants was as follows. . . . For each HUF line derivation, the adult donor was consented and an inner arm 4 mm skin punch biopsy was obtained . . . by a qualified dermatologist.[82]

The fact that CIRM addressed the procurement of non–embryo or gamete derived human donor tissue at all was unusual. For example, at this time, papers from all over the Anglophone world tended to mention the type of tissue from which the iPSC lines were derived (such as human fibroblasts from adults or from juvenile foreskin) and the companies from which cells are bought, but not say anything at all about the procurement or provenance of these cells, despite acknowledgments of intellectual property and funding credit.[83] Again, this was true even when the somatic cells were derived from aborted fetuses, despite the implication of this source in the abortion debate, and even though some of the companies in question acknowledge that procurement standards change over time.[84] ATCC's website, for example, contains this comment in its ethics section: "As requirements for procurement of human biospecimens have changed through the years, ATCC's policies also have evolved to ensure that these standards are met."[85]

Is it the case that the need for attention to the procurement, provenance, and disposition of pluripotent cells in iPSC research disappears because eggs and embryos are no longer required? Buying and selling cells lines from foreskin or liposuction (common sources of normal human dermal fibroblasts) raises questions about who profits, who is implicated, and what happens if privacy of cell lines, through advances in bio-informatics, is compromised. It will be a significant missed opportunity if the iPS revolution closes the door opened by hESC research on the ethics of procurement, provenance, and disposition of human tissue.

As a final example of an alternative stem cell bioscape at this time in the United States, I turn to the question of the potential for "dual use" of stem cell research. In 2007, the National Science Advisory Board for Bio-security raised the issue of "dual use" in biological research:

An issue garnering increased attention is the potential for life sciences research intended to enhance scientific understanding and public health to generate results that could be misused to advance biological weapon effectiveness. Such research has been called "dual-use" research because of its applicability to both biological countermeasures and biological weapons.[86]

Did human embryonic stem cell research come into being with a post-9/11 "ontology of the enemy"?[87] The national-security state has long leveraged the dual-use potential of biomedical technologies, with the help of an epistemological legacy of secrecy in the relations between the military and science and technology in the United States.[88] Is stem cell research being weaponized? If so, who is its wielder and who its enemy victim; who might it protect, and who can be collateral damage?[89] Although human pluripotent stem cell research is very young, stem cell science itself dates back to the 1960s. Two Canadian scientists, James Till and Ernest McCullogh, are generally credited with first discovering and characterizing stem cells when they isolated hematopoietic (blood-forming) stem cells in mouse bone marrow.[90] McCulloch and Till's work grew out of experiments carried out during World War II in which mice previously exposed to radiation were able to recover from radiation injury after receiving a transplant of healthy bone marrow cells. Particular cells in the bone marrow—stem cells—produced this radio-protective effect. From Till's and McCulloch's work to the stem cell research of today there is link between the technologies of death that posed the most pressing moral issues in the twentieth century—nuclear technologies—and the technologies of life such as stem cell research that concern us (as nuclear technologies continue to do) early in the twenty-first century. Among prominent clinical stem cell researchers, there are those whose training lies primarily in reproductive and developmental and embryological sciences and those whose work comes from hematological, neurological, and oncological groundings. For example, Irving Weissman, credited with isolating the first human stem cells, exemplifies the hematology and oncology route, and its links to the older irradiation work and its own entrepreneurial activity and biotechnology start-ups.

Perhaps not surprisingly, human stem cell research is not as far from the necropolitics of Cold War nuclear survivability as might at first appear from the civilian, political, and ethical or religious contests over its regulation that characterized the competing bioscapes of the public stem cell debate during the time period of this book. Even those parts of biomedicine most closely identified with civilian life still take shape in part from national— and, increasingly, global—security concerns. Influences run from military funding for basic research, to infectious disease control, to efforts to obtain global competitive advantage, to domestic and international efforts to thwart rogue state and non-state actors positioned as actual or potential enemies. For stem cell research, a kind of dual-use gradient can be witnessed as one moves (metaphorically or literally) across a U.S. research campus from stem cell research in molecular and cell biology to some of those in bioengineering, and from sociology and anthropology of science to political science and public policy. But these aspects are the visible face of dual use. While the California Institute for Regenerative Medicine was shaping up as a case study of translational innovation in response to funding restrictions on hESC at the federal level, an entirely other kind of initiative was progressing apace.[91] Away from the publicity of the stem cell debate over lives, stem cell research was also receiving support from the federal government as part of its beefed-up post-9/11 biodefense programs.[92]

In 2007, an inter-agency report titled *Advancing Tissue Science and Engineering: A Foundation for the Future* was released.[93] It dealt with the Multi-Agency Tissue Engineering Science (MATES) initiative of the Subcommittee on Biotechnology of the National Science and Technology Council, a subcommittee that had been formed to allow the different federal agencies involved in regenerative medicine and tissue engineering to coordinate with one another. The thirteen agencies involved in MATES include the Defense Advanced Research Projects Agency, the Department of Energy, the Environmental Protection Agency, the Food and Drug Administration, the National Aeronautics and Space Administration, the National Institutes of Health, the Office of Science and Technology Policy, and the National

Science Foundation. The report combined goals from regenerative medicine, from human-subjects reform, and from biodefense into one federal strategic plan. The interagency working group that produced the report worked during the period in question in this book, the period of George W. Bush's presidency. An ethical and political debate about human embryonic stem cell research was going on in the national spotlight, while in the background a revolution in regenerative medicine and tissue engineering was quietly being institutionalized. Although the stem cells discussed in this period are envisaged as adult stem cells, the debate itself is notably absent. Yet, just as much as the snowflake bioscape or the pro-cures bioscape, the MATES document presents a bioscape where lives are to be saved from disease or testing, and protected from the enemy as a strategic initiative. These intertwined goals are not premised on private-sector funding, as federal sources of funding are presupposed. Pursuing the goals will not so much produce cures and regenerate the local economy, as make the nation stronger, in research capacity, in health, and in defense. It responds to a necroscape of war in which soldiers lose limbs and soldiers and civilians alike are subject to chemical and biological attack of unknown origin or target. This different bioscape is summarized in the report's four now suddenly connected questions:

What if lost limbs or organs could be regenerated?
What if drug development and chemical testing were more efficient and less costly?
Can deadly biological and/or chemical agents be rapidly identified?
Are there ways to reduce our reliance on animal testing?[94]

The beginning of the post-9/11 period corresponded almost exactly with the beginning of human embryonic stem cell research in the United States. This period has proved dramatically different from the Cold War in many ways, most notably in replacing a single, superpower enemy with the harder to pin down idea of terrorism. Nonetheless, science and technology have remained at the heart of the U.S. national-security state, and it is reasonable to think that this will continue as stem cell research progresses.[95] The ethics of research need to take this into account.

Why was it that in practice these four alternative biopolitical conversations were not driving the ethics of the beginning of human pluripotent stem cell research, despite the serious consequences of ignoring these issues? There were a number of reasons. One major reason was that these issues did not fit well with the embryo-destruction issue, and raising them was taken as a sign that one was anti-science.[96] In some cases, the wrong people, or only some of the right people were at the table; in other cases, the two entrenched "sides" of the issue stymied efforts to change the terms of debate. Even more generically, procedures and institutional roles put in place in the past determine what issues are subject to debate in what forum, and by whom, sometimes sidelining other issues and sometimes actively preventing other issues from coming under consideration. Finally, there are tremendous difficulties in thinking at the same time about things that happen in different domains and on different scales and in different places and with different histories, even when thinking about different things together is important.

One cannot simply note the reasons for conversations not happening and then move beyond those constraints. The commitments represented by each set of reasons for a conversation not happening are themselves powerful parts of the fields in question, and so cannot easily be jettisoned; "frames" or "bioscapes" are not exogenous devices but constitutive technical, political, and ethical trajectories of ideas, bureaucracy, and equipment. There is no way to have the conversations that are currently missing or less audible or consequential than they should be that does not tease out those histories that inform the current state of affairs. Finding shared commitments, recognizing the specificity of differences over time, place, and group, and introducing new mediators are all necessary steps. So is the will to stage and create a hearing for such ethical excavations. Research universities, with different kinds of experts working in close proximity, with large populations of (in theory, at least) relatively representative young people, and with (in theory, at least) a public mandate to think without undue coercion, should be leaders in this regard. They have to struggle against limits of budget and imagination, against power differentials, and against the entropy of the old disciplines, however. If the problem facing my field of science and technology studies used to be that of how to take account of non-human agency, it is now how to enable any kind of agency that is not already an entrenched, capitalized, and material "interest" bundle of humans and non-humans in a given domain. The beginning of pluripotent stem cell research represented a moment for some ethical choreography; the end of the beginning may make alternate bioscapes harder than ever to bring to salience.

PRESIDENTIAL ADVISORS

In the United States since the beginning of human pluripotent stem cell research, three consecutive presidents, Bill Clinton, George W. Bush, and Barack Obama, convened bioethics advisory bodies to address bioethical issues, including human pluripotent stem cell research. Clinton's 1996–2001 National Bioethics Advisory Commission issued the first report on human embryonic stem cell research in 1999, only a year after the publication of James Thomson's breakthrough paper, "Embryonic stem cell lines derived from human blastocysts."[97] The NBAC's report used more modest language to describe the "hope of new cures for debilitating and even fatal illness" than the pro-cures rhetoric that would emerge a few years later, and put it in the context of "the national debate about the ethics of research involving human embryos" and, unlike the pro-curial bioscape, "cadaveric fetal material." This report recognized the importance of the source or provenance of stem cells, and made its recommendations for procurement on these grounds. Notably, it offered a clue as to why the issue of using aborted fetuses would fade from view on both sides of the argument in the coming years, paving the way for a nearly "business as usual" accommodation of iPSC research. The report successfully put fetal cadaveric tissue in the same category as other tissue from dead bodies, and represented the regulatory environment as already fixed and working in that arena:

The use of cadaveric fetal tissue to derive EG cell lines—like other uses of tissues or organs from dead bodies—is generally the most accepted, provided that the research complies with the system of public safeguards and oversight already in place for such scientific inquiry.[98]

The President's Council on Bioethics, which advised President Bush from 2001 to 2008, both reflected and was in part responsible for working out the competing terms of the embryo debate (which I characterized above as "snowflake" versus "pro-curial") within stem cell research.[99] Like the Discovery Institute, Focus on the Family, and the American Enterprise Institute, the President's Council on Bioethics weighed the costs to "human dignity" of allowing the ends of stem cell research to justify any intentional loss of prenatal life if any more *in vitro* human embryos were sacrificed with public funding.[100] It was widely regarded as providing a forum for genuine reflection about issues that many agreed required deep thought, regardless of one's religious or political beliefs.[101]

In June 2009, half a year into his first year in office, President Obama wrote to inform the members of the President's Council on Bioethics that the council had been disbanded. An article in the pro-life media characterized the announcement as "lamentable, if for no other reason than that they will no longer be producing rich works of philosophical reflection."[102] While agreeing that the President's Council on Bioethics was ideologically biased "in favor of dignity and against giving free reign to technological innovations," the author thought that it was likely to be replaced by a new commission that would also be "ideologically biased, though likely in a more narrow way that is in line with mainstream bioethics," thereby refusing to attribute more or less ideology to one side or the other. Both *Nature*'s news blog and the Federation of American Scientists referred in their coverage of the Council's disbanding to past accusations that the Council had favored ideology over science. While both sources seemed to be willing a science opposed to ideology into being as the "gold standard" for the next national advisory panel on bioethics, the Federation of American Scientists seemed to see only Republicans as ideological, while the U.K.-edited *Nature*'s news blog was less ready to assume that a Democratic U.S. president's appointees would be scientific as opposed to ideological: "It remains to be seen whether Obama will fall prey to the same trap in selecting council members. . . . One can, however, reasonably expect . . . a leading bioethicist at the Center for American Progress, the Washington-based think tank that has served as a farm system for appointees in the new administration [to be involved]."[103]

In November 2009, President Obama created a new bioethics advisory body by executive order. It was no longer a council, but once again a commission, and it was no longer working on bioethics; it was now working on *bioethical issues*. The name "Presidential Commission for the Study of Bioethical Issues" indicated the desire of the Obama administration to move toward practical, reason-governed ethical progress and away from the more philosophical focus on human dignity and the moral status of the human embryo. The members of the PCSBI were chosen for their leadership at research universities and in bioethics, and the new committee notably brought deliberative democratic ethics, based closely on theories of deliberative democracy, to the table. In particular, Obama appointed Amy Gutmann to chair the commission. Gutmann, a political scientist and the president of the University of Pennsylvania, brought to the commission her expertise in ethics and public policy, and in deliberative democracy theory and in its application to bioethics, as well as her experience leading a major research university.[104]

Is deliberative democratic ethics up to the challenges of bringing into focus the alternative bioscapes somewhat obscured in the pro-curial stem cell ethics that was the California response to limits on federal funding for stem cell research?

Theories of deliberative democracy criticize aggregative or majoritarian theories of democracy on the grounds that they overemphasize voting and individualist ideas of the self and underemphasize the substantive aspects of democracy, including the moral dimensions of legitimacy and of collective rights and interests, the value of process, participation, and dissent, and the dynamics of consensus building.[105] Proponents of deliberative democracy make a case for the importance of each member of society having a chance to participate in the making of the choices before us in a democratic polity, not just in voting them up or down. Ways of ensuring that each person has a voice, and that no voices are weighed more heavily than others, as well as ways of transcending being dupes to one's own narrow economic interests, are all envisaged. Deliberative democracy seems to respond to the needs of pluralist liberal democracies to mean something substantive by "democracy," including dissent, multi-vocality, and the process of reasoning in the formal political apparatus. Not surprisingly, then, it seems especially fitting for the realm of the life and biomedical sciences and their bioethics.[106] Several European countries carry out "consultations" and other kinds of public engagement with bioethical topics.[107] These approaches have received mixed reviews as to whether they genuinely represent a substantive improvement in democracy around bioethics.[108] In the United States, with a less consultative, more expert and interest driven tradition, and less of a sense of a public to engage, the greatest need of all is for a range of individuals, however qualified, to imagine different scenarios in which different interests arise in conjunction with the same technologies, and then to ask how that would change the allocation of resources, regulation, and the setting of research priorities.[109]

The Presidential Commission for the Study of Bioethical Issues soon began producing important, frame-changing reports, showing the potential of deliberative democratic ethics not only to certify the ethical but also to call out the unethical.[110] For deliberative democratic bioethics to bring into focus the suppressed bioscapes described above, the following challenges must be met:

The ultimate goal is still to produce consensus or majority recommendations, and this limits the extent to which difference can be expressed and sustained, making the protection of minority rights and interests difficult.

The arbiter of deliberative democracy is reason, and this limits the extent to which those unable to exercise the faculty of reason can participate qua themselves, raising considerable issues for disability justice.

Insofar as the principles of fairness or equal representation are sustained by the ability to imagine oneself in someone else's shoes, there is no basis for acting when empathy cannot be presumed, such as between human and non-human, or between those with whom we radically disagree or who we hate.

Proponents of deliberative democracy have envisaged ways in which some of these challenges can be met,[111] but it is unrealistic to think that deliberative democratic ethics is a panacea. It will remain necessary to seek to perform the ethical choreography of actively seeking to bring to vitality alternative ethical landscapes.

THE ETHICAL CHOREOGRAPHY OF GOOD SCIENCE

After a journey through science with ethics, the snowflake and pro-curial landscapes of early human pluripotent stem cell research in the United States, alternative bioscapes and necroscapes, and the strengths and limits of deliberative democratic bioethics, I now return to the question "What is 'good science' in the age of sciences with ethics?" The expression "good science" commonly refers to science that makes a decisive difference to and/or opens up a field of research. It can also refer to science that is carried out in line with established procedures that promote such goals as institutional and methodological transparency and accountability, integrity of researchers and research protocols, and replicable results—procedures that are embedded in science education and the management of research institutions. Both of these kinds of good science invoke scientific understanding and technical excellence in the field in question; they also both rely on social norms, judgments, and arrangements of many kinds, including committing resources to promising fields of inquiry and positioning a research program in relation to the published literature in the field. They both also demand complex moral commitments as well as scientific reasoning from scientists, such as the desire to devote a life to science, or a compulsion to address a particular problem with fundamental scientific and/or real world significance; to make a real difference. My research made it evident that even—or especially—when scientists are focused on their research, scientists are acting as part of the moral fabric of the stem cell research.[112]

The "good" in these first two uses of the expression "good science" is in some sense internal to the science itself, in the first case to the scientific value of the work done and in the second to the research environment and people's conduct therein. Saying that a great piece of science or science conducted with research integrity evokes the good in "internal" ways does not mean that the good is technical rather than social, for it is surely and deeply both; rather, it means that what is good about the science has to do with judgments about scientific conduct and scientific research itself. These kinds of good science are the *sine qua non* of a flourishing research environment. It is possible for there to be lapses in the kinds of institutional and individual integrity upon which these first two kinds of good science are premised. Prominent examples of scientific fraud and misconduct come to light with some regularity, sometimes provoking widespread scrutiny of peer review and the ethical education of scientists, which in turn tends to bring to light worryingly systematic lapses from protocol and pressures to succeed that can contribute to fraud. However, ideas of fraud and misconduct make sense only against a background where there is the possibility and the presumption of an absence of misconduct and fraud, and the detection of wrongdoing supports the norm and the aspiration of both these kinds of good science. In common with many other fields, the hierarchies in science are significant and the potential rewards are both great and dependent on the work achieved (and getting credit for it). There is thus a fungibility of status and scientific excellence: the desire to solve a "hot" scientific problem, for example, can become a desire to enrich or empower oneself because the two walk—perhaps leap and plunge—in tandem. In the remainder of this book, I will often refer to excellent science, and will credit scientists responsible for that work. I will also refer to a notable instance of fraud and misconduct (though I resist the idea that it is fundamentally disjunctive with ordinary science or with what was going on in other parts of stem cell research at the time). But, as I indicated at the outset of this chapter, the two meanings of "good science" I have rehearsed so far are *not* the meanings of the expression that are of interest to me in this book. I am, above all, *not* concerned with the conduct of any particular scientists or with scientists in general, or with any kind of judgment of them. Likewise, I have no interest in either pushing or reviling science. Praise or blame and pushing or reviling science are data in my inquiry, and not analytic categories, as when a patient advocate promotes science in stronger and less equivocal terms than scientists themselves or when a government berates itself for fraud committed by its nationals. I have spent time with many physicians and scientists—as well as with many humanistic, social-scientific, legal, and activist experts—during the research for this book; I am no more a judge than any other individual, but as far as I could tell these various kinds of experts are individually neither less nor more ethical than people in any other walk of life, if such a comparison even makes sense, and this book has nothing to say about the morality of individual scientists and scholars.[113]

If the first two meanings of "good science" are essential to the modern research enterprise but not those under scrutiny in this book, what then might the expression mean here? For some fields of scientific practice, the word "good" added to "science" invokes an evolving regulatory ethical apparatus that signals that historically fraught societal issues are at stake. This kind of science ethics is more external than the first two not because it is social rather than technical (it too is both) but because the concerns under ethical regulation traverse different domains only some of which are those of scientific and technical practice. Usually this kind of ethical apparatus is the result of, and in turn shapes, negotiations between conflicting domains, and serves primarily to enable materiel (scientific objects, people, equipment, bureaucracy), ideas, and consequences to travel among domains not all of which are primarily scientific. This kind of ethics is common in the fields of bioethics and medical ethics, where the domains of humanity, health, religion, and science (to name but a few) all contend for inter-twined meanings of life, death, the individual and the collective, suffering, sacrifice, flourishing, and the good.

This third class of "good science" has commandeered the word "ethics" for its use (note the slippage in my own usage in the preceding paragraph from "good" to "ethics" and back). On the face of it, this use of the word "ethics" is surprising: the kinds of ethical apparatus in question seem far from the Aristotelian roots of ethics as the study of the quest for human virtue. Bioethics, in my encounters with it (reading and working), is primarily procedural surveillance, and the contemplation of the highest good or the cultivation of virtue through the practical and contextualized pursuit of that which is undertaken for its own sake is at most a backdrop. Similarly, this kind of science ethics hardly looks Kantian: there is little or no individual action and usually no deontological principle at work. Indeed, bioethics seems almost to render redundant precisely those dimensions of the ethical that require reference to behavior, virtue, training, the acceptability of a means or an end, or a greater good or good unto itself. In the United States, the labor of bioethics is typically distributed among people tapped for their distinct expertise, and to an institutional ethics committee on which they all serve. Likewise, the work of the committee typically involves regulatory compliance rather than ethical deliberation, making sure that regulations enacted at various governmental and/or institutional levels are followed. While some more wide-ranging issues can occasionally be raised in these forums, it is usually a misrecognition of one's role to attempt to do so. Serving on these committees, one has the impression that the ethical work has already been done when the regulations were negotiated, or even that the ethics itself has somehow been bypassed. Frequently, this impression is heightened by an institutional impatience that views the work of these committees as a regulatory burden that is at best a necessary evil and at worst a sign of pampering to ignorant social or political interests that serve only to hamper the progress of innovation. When I first began serving on embryonic stem cell research oversight committees, it took me a while to understand that the topics that I was teaching to our students as the principal investigator for the Project on Stem Cells and Society at our Stem Cell Center were rarely going to be germane to task of oversight of a research ethics committee. In 2005, I heard a feminist legal historian, who was summarizing the debates that had been raised during the run-up to the election for and against California's Proposition 71 in 2004, comment that "bioethics means tell me what you [the doctor or scientist] want to do and I'll tell you how you can legally and ethically do it." Her point was that bioethics was, if anything, the opposite of open-ended ethical inquiry; that it was the practice of institutionalizing acceptable ways around people's ethical objections. I watched, and to some extent participated in, the laborious process of working on regulations with whose compliance stem cell ethics committees are charged. Though there is a lot of compromising, real work is done. Putting these regulations into action, then, is first and foremost about enabling research in an environment of ethical controversy, and not about ethical inquiry. But the regulations in question do embody ethical labor of a serious kind, and the kinds of regulations that have to be in place, such as informed consent, recapitulate a distinctive and hard won history of bioethics.[114] This third kind of "good science" is *descriptive* of much of what is being discussed in this book: science that has ethics. Human pluripotent stem cell research in the United States is science that has ethics, and which thus cannot proceed without this kind of ethical accounting. This kind of good science, though, is not the same thing as what I am advocating by calling in this book for good science.

In this chapter I have been advocating "good science" in a fourth sense of the term. After two decades of research and participation in rapidly changing arenas of the biomedical life sciences—stem cell research, reproductive technologies, genomics, and synthetic biology, and associated debates about health disparities, disability justice, race-based medicine, abortion, markets in bodies and body parts, bio-banks, medical tourism, animal rights, intellectual property, philanthropy, and innovation in development and national security—I have become a passionate advocate for pursuing simultaneously the best science and the best ethical practice conceived not merely as overcoming ethical barriers to research, but as articulating the changing biopolitical and necropolitical landscapes of fields of research, and situating funding, discovery, interpretation, and dissemination of scientific knowledge in relation to a multi-vocal mandate to make live, rather than die. I take this to mean, at the very least, that dissent and assent and other interests in relation to fields of science should be solicited, not shut down by scientists and ethicists and administrators; that criticism of science should open up, rather than shutting down avenues of research;

that the process and procedures of ethical inquiry should be honored; and that multiple forums for ethical deliberation should be developed, recognized, and made integral to robust science. The kinds of energy and level of engagement of sciences that have ethics are exemplary; ethical and scientific innovation around stem cell research in its first dozen years was extraordinary and involved much more than downstream implications of the science. However, this energy came from, and the work overwhelmingly reflected responses to, the question of destroying embryos.[115] How could this level of attention be devoted to much more fully vitalized ethical landscapes, to drawing out the evolving and multiply centered biopolitical and necropolitical landscapes of this (or any other) field of research, including but not limited to the question of the embryo? New domains have to be connected one to another, and intellectual and economic capital must come to depend upon this happening. I am asking a lot of scientists in asking them to refuse the view that the public is simply ignorant and welcome dissent and feedback and complication; I am also asking a lot of ethicists by suggesting that shutting down certain kinds of science is a less ethically thorough path than opening up the ethical and scientific debate and insisting that solutions are forged together that inherit specific histories, are revised continuously, recognize as many interests as there are and actively and open-endedly solicit new interests, and do the ethical and scientific choreography of integration. It will take an enormous amount of work, but I believe that something closer to this kind of good science is possible before the field becomes sufficiently normalized for the ethical attention to dissipate and the intertwining of justice and truth to fade yet again into the background. The pro-curial frame, despite being overly based on a reaction to abortion politics, at the expense of other major ethical issues especially issues of distributive justice, and despite turning being pro-cures into being pro-biotech business rather than pro-socially robust solutions to illness and disability, at least had to pay attention to tracking body parts and accounting for funded research in which human body parts are used. It will be a true ethical loss if these advantages are not somehow maintained as this research moves forward.

II STEM CELL GEOPOLITICS

The year 1998 saw the publication of the first scientific paper announcing the successful derivation of human embryonic stem cell lines.[1] James Thomson and his colleagues at the Wisconsin Regional Primate Research Center wrote "Human blastocyst-derived, pluripotent cell lines are described that . . . after undifferentiated proliferation in vitro . . . maintained the developmental potential to form . . . all three embryonic germ layers." In announcing this close-of-century achievement, they speculated that "these cell lines should be useful in human developmental biology, drug discovery, and transplantation medicine." To some, this use of scientific deadpan resonated with the most famous scientific understatement of all time, James Watson and Francis Crick's mid-century masterpiece of the passive voice upon the decoding of DNA's double helix: "It has not escaped our notice that the specific paring we have postulated immediately suggests a possible copying mechanism for the genetic material."[2] Half a century into the greatest life-sciences story on Earth, it was the beginning of the beginning of a subplot: human embryonic stem cell research (hESC).

James Thomson was the American face of human embryonic stem cell research in those early years. A professor of anatomy at the University of Wisconsin-Madison Medical School and the chief pathologist at the Wisconsin Regional Primate Research Center, Thomson had had doctoral training in both veterinary medicine and molecular biology at the University of Pennsylvania and had carried out his dissertation research on the role of genetic imprinting in mammalian early development at the Wistar Institute in Philadelphia. He had spent two years as a post-doctoral research fellow in the Primate In Vitro Fertilization and Experimental Embryology

Laboratory at the Oregon Regional Primate Research Center, then had moved on to the University of Wisconsin at Madison. While in Madison, he also did work in primate *in vitro* fertilization and embryology at the Wisconsin Regional Primate Research Center—work that culminated in the derivation of the first primate embryonic stem cell lines in 1995 and then the first human embryonic stem cell lines in 1998.

The significance of the derivation of human embryonic stem cell lines lay in their ability to proliferate while remaining in a pluripotent state, without differentiating into the specific tissue types that make up the different parts of the human body. In principle, under the right lab or clinical conditions embryonic stem cells could be directed to begin to differentiate so as to become whichever type of cell is needed. The therapeutic potential of these cells was apparent from the start: in theory, these cells could be raw materials from which replacements cells for any part of the body could be induced to grow. Thomson's work was equally important for the development of tools and techniques for isolating embryonic stem cells, and for deriving and maintaining lines and directing subsequent differentiation. For example, Thomson's group developed transfection techniques, methods of homologous recombination, and culture conditions that made it possible to derive and stably maintain human embryonic stem cells. Thomson's group and others went on to show the potential of embryonic stem cells to differentiate in lineage-specific ways, a necessary step in translational embryonic stem therapy. In 2007, he was part of one of the two groups that announced success in human induced pluripotent stem cells, a year after Japan's Shinya Yamanaka and colleagues had published a paper in *Cell* showing the same feat using mouse fibroblasts.[3] The subsequent merging of Thomson's three University of Wisconsin spin-off companies, Cellular Dynamics, Stem Cell Products, and iPS Cells, into the single Cellular Dynamics International (CDI) in November 2008, began the era of the industrialization of basic stem cell technology in the United States.

The possibility of inducing pluripotency in human somatic cells (and thereby having one way to get pluripotency without using women's eggs or human embryos as a raw material, though these continued to be very important in practice) signaled that the first salvo of human pluripotent stem cell research in the U.S. was over. Barack Obama's election as president in the fall of 2008, and his reversal of George W. Bush's stem cell policy in July 2009, did not by any means silence the previous sources of dissent (lawsuits continued to make their way through the courts), but it did re-embed pluripotent stem cell research in the federally funded national research enterprise.

In the autumn of 2001, I was engaged in what must have been one of many early attempts around the country and the world to translate the promise and challenges of human embryonic stem cell research to humanities and social-science classrooms. President Bush's policy was only weeks old, and I was in the midst of interviewing more than 130 Harvard undergraduates for places in a twelve-person freshman seminar on stem cells, reproductive technologies, and the history of bioethics. "What," I asked, "interests you in stem cell research? Is it the scientific potential, or the ethical, religious, and political debate, or the history of bioethics or the life sciences?" Some wanted to be doctors or scientists. Others could not see why anyone would be so ridiculous as to pit the "life" of a few cells in a dish against the life of someone suffering from a disease that might with research be cured, and they wanted to educate and persuade. Others knew that their coming to Harvard had just boosted the price of their eggs greatly, and still others knew something of Nazi science. One brave soul even thought that society ought to be able to, at least in principle, sometimes say no to science; that science was a kind of addiction. Then, quiet at first but quickly escalating, there was a second sound track; cries in the corridor, rising to shouts, and all around opening doors. We were out of our offices in a matter of moments, joining the press of people getting down the stairs somehow and then watching in horror the looping footage on the large screen in the auditorium on the ground floor of the Science Center. The Twin Towers had been hit. Bodies and disintegrating edifices fell, fast and slow, again and again. The seminar selection process in suspended animation, weeping students instead used my office phone as they tried to contact family members. Each time we met over the semester, the new technologies and politics of death cast their shadow into our classroom on these new technologies and politics of life.

In the fall of 2003 I moved to California, where I began almost immediately receiving mail and email inviting me to join in supporting the upcoming Proposition 71, the California Research and Cures Act. I also quickly made the acquaintance of various groups who supported embryonic stem cell research in general but were opposing the proposition either on the grounds that it was fiscally irresponsible or on the grounds that it was likely to be exploitative of women or to take us into a brave new world of cloning, germ-line genetic modification, eugenics, and chimeras. California exceptionalism and a Pacific rim and Americas orientation, as well as distinctive immigration politics and a culture of patient and celebrity disease advocacy, pummeled my senses with the biopolitical and necropolitical differences between the coasts. I had come to the stem cell state to teach and research and live; life and death were the same yet very different.

In chapter 2, I described how the ethics of stem cell research got bundled with particular constituents and institutions and forms of material practice at the beginning of human pluripotent stem cell research in the United States. I discussed the ways in which ethical dissent was operationalized around procurement policies that specified acceptable derivation. I focused in particular on what I called the "pro-curial" solution to permissible stem cell research that emerged in California in response to federal policy, and I discussed the ethical concerns that were central to this solution and some of the ones that were more and less sidelined by it. I argued for an active ethical choreography (requiring the linking together of historical precedent, material culture, and ethical perspective) to, as it were, re-wire or re-set the visibility of concerns to do with distributive justice and inequality, disability justice, and dual-use applications.

Having focused on how hESC was turned into doable science with ethics in chapter 2, in this chapter I look at how different world views containing hESC were crafted. I do this in two steps. First, to excavate the ways that Republican and Democratic views on stem cells differed even while they evolved in relation to one another, I compare and contrast President Bush's and President Obama's speeches announcing their respective stem cell policies. Second, I examine certain striking aspects of the language of Proposition 71, and some responses to that. I focus on my own attempts to develop a curriculum for CIRM-funded students that went beyond the ethical frame of Proposition 71, and on a thought experiment upon which I found myself embarked. As I explained in chapter 1, it is important to the narrative arc of this book that Proposition 71 and responses to it by local civil-society groups and activists (as well as a renewed debate about national competitiveness vis-à-vis other countries with fewer funding restrictions) occurred between the two presidential speeches in time, as a result of the federal funding limits imposed by Bush.

In order to make the comparisons at stake in this chapter as clear as possible, however, I will first compare the presidential speeches and then address the California debate.

BUSH'S "FUNDAMENTAL QUESTIONS" VERSUS OBAMA'S "RESPONSIBLE . . . CONSENSUS"

In a landmark speech delivered on the evening of August 9, 2001, President George W. Bush addressed the nation from his ranch in Crawford, Texas. His topic was "research involving stem cells derived from human embryos."[4] The choice of time and locale suggested that he was speaking of a topic that did not primarily concern the working day, Washington, and business as usual; instead, the president wanted to "discuss with you" a topic that was both a "national debate and a dinner table conversation," that involved heteronormative nuclear families ("parents," "couples," and "children") as well as experts, and that required the searching of one's conscience about an issue that was "one of the most profound of our time."

President Obama's speech of Monday, March 9, 2009, consisting of remarks prepared for the signing of an executive order titled Removing Barriers to Responsible Scientific Research Involving Human Stem Cells, and the simultaneously introduced Scientific Integrity Presidential Memorandum, began very differently.[5] "We" (the word Obama used as the subject of the first sentence) were about to sign an executive order that would "bring the change" "hoped for, and fought for," by "scientists and researchers; doctors and innovators; patients and loved ones." In the place of a discussion with the whole nation around their dinner tables, the 44th president was taking executive action on behalf of stem cell research's supporters, in the core of his working day, to "vigorously support scientists" and for "America to lead the world in the discoveries it may one day yield."

The speeches went on as differently as they started. Although both presidents relayed scientists' belief in the promise of embryonic stem cell research (Bush: "scientists believe further research using stem cells offers great promise that could help improve the lives of those who suffer from many terrible diseases"; Obama: "scientists believe these tiny cells may have the potential to help us understand, and possibly cure, some of our most devastating diseases and conditions"), both proclaimed their faith (Bush: "I

. . . believe human life is a sacred gift from our creator"; Obama: "As a person of faith"), and both acknowledged that Americans differ in their beliefs on the matter (Bush: "people of different faiths—even many of the same faith—coming to different conclusions"; Obama: "Many thoughtful and decent people are conflicted about, or strongly oppose, this research. . . . But . . . the majority of Americans . . . have come to a consensus that we should pursue this research"), the differences in how they expressed these similarities, and in the content and import of their speeches in other regards, were striking.

For the empirical study of ethics,[6] it is notable that the meta-ethical positions adopted in the two speeches were radically different. In President Bush's speech, his stated ethical position was a deontological one, albeit one with some room for maneuver imparted by the strategic insertion of the little word "any." In weighing the promise of stem cell research against destroying human embryos, Bush said that "even the most noble ends do not justify any means." This statement departs only slightly from the familiar deontological slogan "The end does not justify the means." Bush's version could imply either that the "noble ends" don't "justify any means" *ever* or that they don't justify each and every (or any old) means. In other words, common usage grants a certain ambiguity to a position that could be understood as absolutist (the end never justifies the means), or merely presumptively Kantian (noble ends do not as a matter of course justify using any means to reach them, so the means needs a lot of thought).[7] The means in question are the destruction of human embryos, but, beyond that, their wanton destruction in such a way that the culture becomes one that "devalues life." For Bush, "this issue forces us to confront fundamental questions about the beginnings of life and the ends of science," "juxtaposing the need to protect life in all its phases with the prospect of saving and improving life in all its stages." The result is a moral dilemma about life.

In Obama's remarks, on the other hand, the stated ethical position was compatibilist as regards science and morality, putting forward the position that there is in fact no fundamental tension between the ends of science and ethics in the field of stem cell research, and that doing science responsibly would resolve any apparent tension.[8] This position that it "is a false choice between sound science and moral values . . . the two are not inconsistent" aimed to make the deontological position of not allowing an "end to justify any means" moot. The Obama administration might have put forward a consequentialist position, arguing that these particular means—using (e.g., leftover IVF) embryos for research—could be justified by the ends of science and the moral imperative to find cures, and indeed the remarks did emphasize the economic and nationalist need for cutting edge science that might go elsewhere if not promoted at home ("we will ensure America's continued global leadership in scientific discoveries and technological breakthroughs," or otherwise other "countries may surge ahead of ours"), but overall the remarks were not about the ends justifying the means.[9] Instead, Obama refused the terms of the deontology-versus-consequentialism debate, confining moral issues to the realm of norms guiding responsible scientific conduct and regulation permitting rigorous oversight of research and research materials. The cost of this position was that Obama left it unclear in what the "difficult and delicate balance" consisted, if science and ethics were compatible. Likewise, although he told the nation that "with proper guidelines and strict oversight, the perils can be avoided," it was not clear to what "the perils" referred. Instead, the target became the distortion of science and reason that the compatibilist position implicitly imputed to any position recognizing the potential for moral conflict between science and morality. In this way, Obama's remarks connected to the Democratic attack on Republicans' alleged disregard for the "inconvenient truths" of science ("ensuring . . . that we make scientific decisions based on facts, not ideology," "especially when it's inconvenient"); that expression, in turn, referenced *An Inconvenient Truth*, a 2006 documentary film about former Democratic Vice President Al Gore's campaign to persuade Americans of the reality of global warming. Accordingly, Obama's remarks were accompanied by the "signing of a Presidential Memorandum . . . to develop a strategy for restoring scientific integrity to government decision making." The battle lines, then, were drawn on positions that were not simply two sides of the same debate. The Democrats, as the Republicans would have it, promoted a culture that devalued the intrinsic worth of each unique "snowflake" life at its beginning; the Republicans, as the Democrats would have it, ignored reason and threatened species life by doing so. The speeches effectively anchored the sides, respectively, to the abortion debate and to the climate change debate, and thus to the larger political stakes for which these issues have come to stand for each party.

The epistemological positions that accompanied these ethical positions might at first seem counter-intuitive, although I argue that it is in the

combination of the epistemological and ethical that the differences between
Republican and Democratic views on human embryonic stem cell research
really reveal themselves.[10] The duties or principles of deontological ethics
(here, don't destroy life in its beginnings) are sometimes criticized for a
surfeit of certainty or absolutism, at the expense of sensitivity to context
and/or the greater moral good.[11] In contrast, ethical positions that aim to
promote particular ends (here, cures from stem cell research and a scientifically vibrant economy) are sometimes faulted for their burden of uncertainty: how can not-yet-actualized ends be known and assessed for their
moral value? Yet it was Bush's speech that was riddled with uncertainty,
attributed to himself ("I pray it is the right" "decision"), to science ("no
one can be certain that the science will live up to the hope it has generated"), and to the people ("Many people are finding that the more they
know about stem cell research, the less certain they are about the right
ethical and moral conclusions"). Obama's remarks, on the other hand, were
brisk (1,219 words, versus Bush's 1,732). Obama apparently felt no need
to "ponder" (Bush) and nothing about which to be "agonized" (Bush).
Instead his remarks were paternalistic and technocratic ("To ensure that in
this new Administration, we base our public policies on the soundest
science; that we appoint scientific advisors based on their credentials and
experience, not their politics or ideology") and suffused with decisiveness
("we will bring the change"; "we will lift the ban on federal funding";
"We will vigorously support scientists who pursue this research"; "we will
aim for America to lead the world in the discoveries it one day may yield";
"We will develop strict guidelines which we will rigorously enforce") and
a majoritarian certainty ("The majority of Americans . . . have come to a
consensus that we should pursue this research").[12] Obama addressed the
"hope and hype" that is reflected in the over-promising of science, but he
pledged to support the effort to realize that promise ("Ultimately, I cannot
guarantee that we will find the treatments and cures we seek. . . . But I
can promise that we will seek them"). He also lauded lives led in the belief
in the ability to progress that goes with hope and promise. To mark the
difference between what can currently be delivered (which should not be
over-hyped) and the moral life of never taking no for an answer (which
cannot be over-hyped), Obama referred to the by-then-deceased actor
Christopher Reeve, who had been left quadriplegic by a horseback-riding
accident and whose gym had a sign that read "For everyone who said, 'It's

impossible.' See you at the finish line." Bush, on the other hand, noted that we had reached the "brave new world" that had once seemed so far off, of science running amok. He also pointed to past failures to deliver on scientific promise as its own kind of morality tale about scientific hubris ("scientists believed fetal tissue research offered great hope for cures and treatments, yet the progress to date has not lived up to its initial expectations"). On Bush's view, those with objections to the excesses of science were the conscience of a society where science is not the only game in town ("I hope we will always be guided by both intellect and heart, by both our capabilities and our conscience"). For Obama, however, the fact that a majority of Americans supported stem cell research by the time of his speech translated into "a consensus" composed of the views of "the majority of Americans." Those disagreeing were entitled to respect, but did not have any role regarding the running of government, and must not stand in the way of science.

As can be seen in table 3.2, a word count shows us these parameters in a slightly different way. For Bush, the difference between the private and public spheres, and their respective private and taxpayer funding bases, were pivotal. The Republican position aimed to circumscribe only what could be done in the public sphere, with tax dollars. The Democratic position, on the other hand, did not differentiate the private sphere from the general public good of science. Bush placed ethics in potential tension with science; Obama did no such thing, mentioning moral values only once. Importantly, both presidents emphasized science, research, promise, hope, and cures (the pro-curial package). Though branded by Democrats as anti-science, Bush characterized himself as "a strong supporter of science and technology" who, "like all Americans," had "great hope for cures." The most striking difference between the speeches involved the words "life" and "work." Bush tied the question of federal funding for hESC research to the ethical issue of life in its early stages; Obama tied it to the economic need to be at the cutting edge of, rather than held back in, the work of science.

In the end, the differences between what could be funded under Bush's policy and after its repeal were arguably less significant (given other federal restrictions, discussed below) than the differences in political world view and the emerging place of hESC research in each world view. For the purposes of this book, the most important conclusion that can be drawn

TABLE 3.1

A thematic comparison of President George W. Bush's August 2001 speech and President Barack Obama's March 2009 remarks.

	Bush's speech	Obama's remarks
The good	"I worry about a culture that devalues life" "Even the most noble ends do not justify any means."	"a false choice between sound science and moral values . . . the capacity and will to pursue this research—and the humanity and conscience to do so responsibly"
Allusion that connects position on stem cell research to political party's broader views	"Snowflake"(allusion to trademarked name of a pro-life Christian frozen embryo adoption and donation program; personhood of human embryos)	"Inconvenient " (allusion to name of former Democratic Vice-President Al Gore's film on climate change skeptics, An Inconvenient Truth; threat to science from ideology)
Epistemology	Science and ethics not always reconcilable; science must not trump ethics; uncertainty	Science and moral values compatible; scientific nation; majoritarian certainty
President's task at hand	"My administration must decide whether to allow federal funds, your tax dollars, to be used for scientific research on stem cells derived from human embryos."	"We will vigorously support scientists who pursue this research. And we will aim for America to lead the world in the discoveries it one day may yield."
Public figure patient advocate	Nancy Reagan (wife of former Republican President Ronald Reagan, who latter suffered from Alzheimer's Disease)	Christopher Reeve (actor who sustained spinal injury in riding accident)

TABLE 3.2

Word counts of President George W. Bush's August 2001 speech and President Barack Obama's March 2009 remarks.

Words used	Bush's speech	Obama's remarks
public, private, tax, taxpayer	4 (0.2%)	1 (0.1%)
ethics, ethical, ethicist, moral	14 (0.8%)	1 (0.1%)
science, scientist, research, cure, promise, hope	64 (3.6%)	40 (3.2%)
work	2 (0.1%)	8 (0.7%)
life	24 (1.3%)	3 (0.3%)
faith, religious, church	4 (0.2%)	1 (0.1%)
Total number of words	1,732	1,219

from this comparison of the two speeches, however, stems from what got clumped together. Uncertainty and asking the general public to think deeply about issues emerged as anti-science, and as allowing government science policy to be driven by ideology and religion rather than reason. Supporting the use of the best possible science in the service of governance, jobs, and international competitiveness emerged in a technocratic vein; a belief in scientific cures emerged as the progressive alternative to ideology. Different combinations of these elements (for example, a desire to combine deep thought with the best possible science, or to seek not only cures but also accommodation, acceptance, and access) were made that much harder to imagine or inhabit. It will take ethical choreography to re-wire these elements if there is to be hope of bringing the under-examined bioscapes discussed in the last chapter more to the fore.

Each of the two speeches accompanied the implementation of specific policies for human embryonic stem cell research funding. It is important to understand, however, that these policies were under-determined (not every aspect of "acceptable derivation" was specified, as was noted in chapter 2), did not necessarily mesh with other sources of applicable federal policy (for example, the Dickey-Wicker Amendment, discussed below), and did not in their particulars capture the views of all members of the respective political parties in the years between the speeches. It is against this dynamic background that transition from stem cell nation to stem cell state—and to some extent back again—occurred.

President Bush's speech significantly restricted, but did not ban, federal funding for human embryonic stem cell research.[13] His decision was that federal dollars for human embryonic stem cell research (the ruling did not affect individual state or private funding for hESC research) should be restricted to research using stem cell lines that had already been derived, so that no new embryos would be destroyed under his watch but research would not be stymied altogether.[14] The lines that were declared at the time of the policy to meet these criteria and were subsequently listed in the NIH registry established in light of the policy became known as the "presidential lines." These lines came not only from laboratories in the United States, but also from India, Israel, Singapore, Sweden, and South Korea. To be a "presidential line," a line also had to have been generated from leftover IVF embryos without financial inducement to the gamete donors. It was still the early days of human embryonic stem cell line derivation—the first lines had been derived in 1998, and problems with the use of these lines for research and eventual therapies rapidly emerged.

The source of embryos for research that Bush considered and rejected for eligibility for federal funding for research was embryos leftover from infertility treatment that were still alive (though frozen). So-called leftover embryos, as has already been noted, are no longer needed for their intended parents' reproductive future, but are not yet dead. Bush committed federal funds to adult stem cell research, and announced the appointment of Leon Kass to run the President's Council on Bioethics. Bush rejected human cloning and the production of humans to supply spare body parts.

Not every Republican agreed with this policy. By 2005, it was apparent that some prominent Republicans were breaking with the Bush policy. For example, William Harrison "Bill" Frist of Tennessee, the Senate's Majority Leader since 2003, began to waver in his support of the president's policy as the viability of the majority of the Bush-approved stem cell lines came into question, as the science itself progressed, and as public and bipartisan support for stem cell research increased. On July 29, 2005, Frist announced in a speech from the Senate floor that he had decided to part ways with the president on the matter of support for hESC research, and he thenceforth became a vocal, knowledgeable, and highly influential supporter of HR810, the Castle-DeGette Stem Cell Research Enhancement Act of

2005 that passed in the House of Representatives, and the corresponding Senate bill that passed on July 18, 2006. Both bills would have made federal funds available for embryonic stem cell research using frozen embryos left over from infertility clinics where both the egg donor and the sperm donor had given their consent and where the embryos would otherwise have been discarded.

A second group of generally pro-life Republicans who broke from Bush's policy was exemplified by Nancy Reagan, the former First Lady of California and of the United States. Her views on stem cell research were those of a passionate patient advocate despite her pro-life, anti-abortion views. Rather than focusing on the *in vitro* embryos that must be destroyed to create embryonic stem cell lines (as do abortion opponents who also oppose embryonic stem cell research), Mrs. Reagan emphasized the good that could come from the cures that might be developed to help those with various diseases and debilitating conditions. She became a particular advocate for the use of embryos that were stored frozen at infertility clinics and were no longer needed for reproductive purposes. Mrs. Reagan's patient advocacy came both from her own struggle with breast cancer and from her care over a decade of her husband, former president Ronald Reagan, prior to his death in 2004 from Alzheimer's Disease. Her conservative voice was important in galvanizing Republican support for stem cell research in a partisan debate. She was also part of a trend in disease advocacy in that she was a celebrity. As Michael J. Fox, Christopher Reeve, and other celebrity advocates of stem cell research had done, she used her resources and public and media recognition to advance the pro-cures cause.

In May 2005, H.R. 810, the Stem Cell Research Enhancement Act of 2005, passed in the House, and a similar bill in the Senate, only to be vetoed by President Bush the following year. In June 2007, Bush again vetoed stem cell legislation that would have freed up federal funding for stem cell research using frozen leftover (or in excess of clinical need) embryos that would otherwise be discarded and for which there was parental consent to donate to research. In its place he issued a new executive order that made two important rhetorical changes.[15] First, the National Institutes of Health's registry of stem cell lines eligible for federal research funds, formerly known as the Human Embryonic Stem Cell Registry, was renamed the Human Pluripotent Stem Cell Registry. Second, the order articulated Bush's deontological principle in a clearer way, refusing the sacrifice of one life for the medical benefit of another (although this executive order was still not quite categorical; it retained the qualifier "mere"). Additionally, the order linked Bush's position on embryo destruction to concerns about the commodification of humans—a less partisan issue because of its appeal to many progressives and many religiously motivated voters—by defining embryos as part of the human species:

(c) the destruction of nascent life for research violates the principle that no life should be used as a mere means for achieving the medical benefit of another;
(d) human embryos and fetuses, as living members of the human species, are not raw materials to be exploited or commodities to be bought and sold.

Over time, the view that Bush's stem cell policy stifled science and innovation gained ground. Senator Hillary Clinton (D–New York), for example, accused Bush of putting "ideology before science."[16] Once the possibility of replacing human-embryo-derived pluripotent stem cells with somatic-cell-derived induced pluripotent stem cells became a reality, efforts began to establish whether in fact the two kinds of stem cells are functionally equivalent and equally safe for therapeutic purposes. Several top stem cell research groups went on to publish work showing that it is not necessary to use a viral vector to transfect the pluripotency inducing transcription factors into the non-pluripotent cell, or to use oncogenes.[17] At the same time, efforts to characterize gene expression in both hESCs and iPS cells seemed to show that some differences remained,[18] which may or may not prove clinically relevant in the longer term. In any case, as scientific expertise increased around the world and as the research became more normalized, popular support for embryonic stem cell research grew. Polls showed that a majority of Americans favored embryonic stem cell research as well as adult stem cell research even at the time of the 2001 announcement.[19] This discrepancy between the president's position and the public meant that Bush's own convictions ended up playing a major role in how stem cell research developed in the United States. While Bush's policy held, several states initiated state-financed stem cell efforts to make up for the shortfall in federal funds and to fill this policy gap. Most people expected, however, that the next administration would act to align federal policy more with popular opinion and lift many of the restrictions on federal funding for stem cell research, especially as the science itself became more advanced. They were not disappointed.

On March 9, 2009, Executive Order 13505, titled Removing Barriers to Responsible Scientific Research Involving Human Stem Cells, was issued by President Obama. The executive order revoked the Bush-era stem cell policy, the policy of August 9, 2001, and Executive Order 13435 of June 20, 2007, and charged the NIH with coming up with guidelines for future federal funding of human pluripotent stem cell research. The Draft National Institutes of Health Guidelines for Human Stem Cell Research described research eligible for funding as first "ethically responsible, scientifically worthy, and conducted in accordance with applicable law."[20] In summary, President Obama's policy change meant the following for federal funding of hESC research: First, the Department of Health and Human Services (including the NIH) could use appropriated funds for research using human embryonic stem cells that were derived from IVF embryos no longer needed for reproductive purposes,[21] but not the derivation of stem cells from these embryos (which remained prohibited by the annual appropriations ban on funding of human embryo research). Like Bush's policy, Obama's policy permitted the funding of research on "adult" stem cells (without addressing the use of "adult" stem cells sourced from fetuses) and on induced pluripotent stem cells. And any research that could be funded had to meet certain conditions and informed-consent procedures during the derivation of the human embryonic stem cells for research using the cells to be funded by the NIH. Federal funds were still not available for hESC derivation, somatic cell nuclear transfer (or research cloning), parthenogenesis, the creation of embryos specifically for research, or anything involving the creation of cybrids and chimeras.[22] The main change, then, was to permit the use of federal tax dollars for research on human embryonic stem cells lines that had been made from leftover IVF embryos, whether or not they had been derived before August 9, 2001.

Although certain important groups with which I worked during this period felt that Obama got his stem cell policy exactly right, this was not the policy for which I advocated.[23] I am wary of the routinization of the use of leftover embryos for research, and have argued that physicians should bring down the numbers of eggs that a woman produces in a cycle of *in vitro* fertilization so as to decrease her exposure to gonadotropins as well as her likelihood of having a multiple pregnancy. After the notorious case of the Los Angeles "Octomom," this need became even more apparent. Having a niche—human embryonic stem cell research—for "leftover embryos" helps confer on them the status as usable and decreases the pressure to bring down the ratio of numbers of eggs harvested to number of pregnancies. I advocated the extremely unpopular position that it was better to ask women to donate eggs (or any donor to donate germ-line or somatic tissue) specifically for research, and then not to alienate the donor from the information or the products or value accruing to that tissue (so my view also went against the bipartisan preference for "donated" eggs and tissue, where alienation of the body materials was considered a concomitant requirement of altruism, which was, in turn, considered a prerequisite in defense against the exploitation of donors). I develop parts of this argument below and in chapter 5, and have written about it in the past.[24]

Obama's action did not address or repeal the so-called Dickey-Wicker Amendment (Title V, Section 509 of House Resolution 1105), which was part of the Omnibus Appropriations Act of 2009, as it had been part of every annual appropriations bill since 1996. Under the Dickey-Wicker Amendment, U.S. government funds cannot be given to research that makes, destroys, discards, or harms any living human embryo. Furthermore, the amendment defines "human embryo" very broadly: "any organism, not protected as a human subject . . . that is derived by fertilization, parthenogenesis, cloning, or any other means from one or more human gametes or human diploid cells." This remained a barrier to using federal funds for research that included creating stem cell lines from embryos and embryo-like organisms. If interpreted strictly, it left it unclear whether federal funding could be used for many kinds of pluripotent stem cell research, including research involving the creation or destruction of embryos and embryoid bodies during the derivation of embryonic or induced pluripotent stem cell lines (for, at the moment of induced pluripotency, might they not meet the definition of "organism" quoted above?).

In a suit originally brought in 2009 by Dr. James Sherley and Dr. Theresa Deisher (both of Nightlight Christian Adoptions) and colleagues against Secretary of Health and Human Services Kathleen Sebelius, the plaintiffs argued that President Obama's lifting of restrictions on hESC funding violated the Dickey-Wicker Amendment.[25] Sherley and Deisher were found in district court to lack standing, but the appeals court found that, as adult stem cell workers fighting for their share of a fixed funding pie, they did have standing.[26] In August 2010, U.S. District Judge Royce Lamberth found the easing of NIH restrictions on funding to be illegal because it violated the Dickey-Wicker Amendment, and granted a preliminary injunction in favor of Sherley and Deisher. The U.S. Court of Appeals for the D.C. Circuit overturned the preliminary injunction, distinguishing between federal funding for work on hESC lines and funding to derive the lines themselves (and thus destroy embryos). Judge Lamberth dismissed the case in July 2011, despite having felt that the Dickey-Wicker Amendment was not ambiguous. The plaintiffs appealed the ruling in September 2011, and in December 2011 a three-judge panel was assigned to the case. All three judges, Chief Judge David Sentelle (appointed by Ronald Reagan), Janice Rogers Brown (appointed by George W. Bush), and Karen LeCraft Henderson (appointed by George H. W. Bush) were conservative, which opened up the possibility of restricting funding more broadly than at any time to date. In August 2012, the U.S. Court of Appeals for the District of Columbia Circuit rejected the case, and in January 2013 the U.S. Supreme Court refused to hear the case, bringing this challenge to a close.

Although this legal challenge was eventually rejected, debates about sciences with ethics do not get resolved once and for all; instead, practice goes ahead through operationalizing points of dispute to bureaucratic and material practices of ethical accounting (for example, through attention to procurement of acceptable cells, for CIRM-funded hESC research, as I argued in chapter 2), and these accounting procedures in turn take their bioscapes, or frames, of who and what counts (the pro-curial frame, for example) from plausible combinations of elements uniting political, economic, and aspirational goals. Once in place, the ethical accounting procedures that make practice doable are hard to undo, as they perhaps should be, even as debate continues. They are also, however, hard to change, including changing in ways that bring different elements to bear on the political, ethical, and scientific questions of how to conduct what kinds of stem cell research.

CALIFORNIA DREAMING (OF REAL ESTATE AND WOMEN)

New Jersey, Wisconsin, Massachusetts, and California were among the states that acted early to fund stem cell research at the state level, to fill gaps in federal funding resulting from President Bush's policy. There were also several states that moved to prohibit all human embryonic stem cell research at the state level, including Arkansas, Iowa, North Dakota, and Michigan, thereby going beyond the Bush policy in the other direction. But California's initiative was by far the most ambitious effort explicitly to channel funds to human embryonic stem cell research. On November 4, 2004, the day Bush was re-elected, Proposition 71, the California Stem Cell Research and Cures Initiative, became California law, by a margin of approximately 59 percent to 41 percent (a landslide in a two-party liberal democracy). The major reasons people gave me for this high level of support were the following:

• the promise that human embryonic stem cell research would one day cure many currently untreatable diseases and disabling medical conditions
• anti-Bush sentiment
• pro-science sentiment
• the insistence of the proposition on funding everything from tissue extraction to clinical trials for therapies
• the promise of regeneration of both medicine and California's economy and fiscal crisis through trickle-down from investment in science and technology
• the need for a supply of human embryonic stem lines beyond the so-called presidential lines because of serious limitations to those cell lines (they represented a small genetic pool, they were contaminated with non-human feeder cells, and several of the lines were non-viable).

Those with whom I talked who had voted against the proposition had done so because of their views on abortion, or because they opposed the proposition on the grounds that it was fiscally irresponsible, or because it proposed a less than transparent governance structure. I also talked to several people who were extremely ambivalent; some of them had wanted to register a pro-science, or an anti-Bush vote, yet had a variety of progressive concerns with the potentially dehumanizing effects of technology, including "gateway" anxiety that stem cell research would license such things as cloning and chimera creation down the line, that it was "playing God," that it was "technology run amok," that it fostered "unrealistic expectations," or that it might lead to exploitation of women who would be needed as egg donors for the somatic cell nuclear transfer process of generating embryonic stem cells that it funded.

As I noted in the previous chapter, the wording of Proposition 71 was crafted meticulously. Its passage added Article XXXV to the California Constitution:

Sect. 4. Article XXXV is added to the California Constitution, to read: . . . There is hereby established a right to conduct stem cell research which includes research involving adult stem cells, cord blood stem cells, *pluripotent* stem cells, and/or progenitor cells. Pluripotent stem cells are cells that are capable of self-renewal, and have broad potential to differentiate into multiple adult cell types. Pluripotent stem cells may be derived from *somatic cell nuclear transfer* or from *surplus products of in vitro fertilization* treatments when such products are donated under appropriate informed consent procedures.[27] (emphases added)

As exemplified in the above paragraph, the text of the proposition used the word "pluripotent" instead of "totipotent" (a practice that I have followed in this book). "Totipotent" was the word that had earlier been used to describe embryonic stem cells.[28] People noted that the inner cell mass (ICM) that had to be extracted from a blastocyst (early embryo) to make an hESC line had the potential to make all the tissues of the body *except* the trophoblast (responsible for making the placenta during pregnancy), and so it did not seem dishonest to claim that the cells were pluripotent rather than totipotent, and it sounded less threatening. The proposition's text also avoided the word "cloning" except where the text explicitly banned "reproductive" cloning. Instead of "research cloning" (the term used to describe the process of taking an egg from a member of a non-human species, enucleating it, and adding the nucleus of a somatic cell), the proposition referred to "somatic cell nuclear transfer" (SCNT) as one source of hESC lines fundable by the passage of the proposition. The only reference to the eggs that would be needed for SCNT was a single mention of the "egg cell" required for the procedure. Like embryos, eggs (plural) were not mentioned at all. The women from whom eggs would be sourced, or the individuals or couples whose embryos would be used, were not mentioned at all either.[29] In short, a constitutional right was established to conduct human embryonic stem cell research in the state of California, and three billion dollars of taxpayer money was pledged to support the research and its real estate requirements, without mentioning women, embryos, eggs, research donors, totipotency, or research cloning.

In early 2005, a statewide competition was held to site the California Institute for Regenerative Medicine (CIRM), the new state agency that Proposition 71's passage had brought into being. San Francisco, thanks to hefty deal sweeteners from the city's government, won. Robert Klein, the Bay Area real estate developer and diabetes advocate who had chaired the Yes on Proposition 71 campaign committee that wrote Proposition 71, became the president of CIRM's governing body, the Independent Citizen's Oversight Committee (ICOC). The ICOC was composed of representatives of groups likely to benefit from stem cell research funding: disease advocates and the leaders of public and private research entities and biotechnology companies. The members of the ICOC met regularly to vote on CIRM's business and to approve the actions of CIRM's working groups. Dr. Zach Hall was elected the first president of CIRM in 2005. In its first year, CIRM had to deal with legal challenges that held up the sale of bonds. Judge Dana Sabraw ruled in March 2006 that a major challenge to CIRM's constitutionality was without grounds; it subsequently cleared the appeals process. In the interim, CIRM members staff and supporters raised private and public bridging funds to tide the agency over until the funds were released.

Over its first few years, CIRM drew up intellectual-property guidelines for non-profit and for-profit research organizations, making access and affordability of therapies developed with CIRM funding a topic of central concern. California Code of Regulations Title 17, Section 100407, titled Access Requirements for Products Developed by For-Profit Grantees, required that for-profit grantees "submit a plan to afford uninsured Californians access to a Drug." They also formally agreed to abide by all extant California laws about diversity and research. Efforts by various lawmakers to improve CIRM's accountability and accessibility and affordability of any products met with varied success. California Senator Deborah Ortiz, who sponsored the bill that legalized human embryonic stem cell research in the state in 2002, fought to tighten CIRM standards to protect egg donors, guarantee affordability of therapies, and broker intellectual-property arrangements that would offer returns to the state. Senate Bill 1260 passed in the fall 2006. But in 2008 Governor Schwarzenegger vetoed Senate Bill 1565, introduced by Democratic Senator Sheila Kuehl and Republican George Runner. The bill had as one of its aims making stem cell therapies and diagnostics funded by CIRM more affordable.

CIRM announced its first Request for Application (RFA) on May 13, 2005. UC Berkeley submitted an application and eventually received one of the training grants that were being funded. The RFA announcement said the following:

Each institution is expected to offer a single, integrated program appropriate for the educational level of its trainees and the expertise of its faculty. All programs will be required to offer at least one course in stem cell biology and disease and a course in the social, legal and ethical implications of stem cell research. The RFA is designed to educate students from scientifically diverse backgrounds—including the relevant fields of biology, clinical training programs, bioengineering, as well as ethics and the law.

I became a principal investigator for the project on stem cells and society at Berkeley's nascent stem cell center, and helped put together and teach the first required ethics classes. Our campus was alone in the state in receiving funding to support a CIRM scholar in the law and one in the social sciences and the humanities as part of our training grant. This was due in part to strength in these fields and a well-organized effort by members of law and social-science faculties early on, but also because Berkeley is alone among major research universities in the state in not having a medical school and so not being in the running for clinical research fellowships, except through a loose partnership with Children's Hospital Research Institute, Oakland. In our work, we argued that the older ethical paradigm, ELSI, was outmoded. (See chapter 1 above.) ELSI, standing for "ethical, legal, and social implications," had been a part of the Human Genome Project (HGP) almost since its inception. In the words of the HGP,

ELSI provides a new approach to scientific research by identifying, analyzing and addressing the ethical, legal and social implications of human genetics research at the same time that the basic science is being studied.

With the passage of Proposition 71, the need for this kind of integrated research went further, we argued, because the social and ethical and legal issues are not downstream "implications" anymore, but are part and parcel of the research itself. The social contract implicit in the corporate, non-profit, academic (both public and private), and governmental chimera that formed CIRM and its grantees required a curriculum that reflected many disciplines and questioned many things typically taken for granted.

PROGRAM ETHICS?

Working with the acronym ELSPETH instead of ELSI, I tried to organize into ethical, legal, social, political, economic, theological, and historical elements the concerns that were emerging while talking and listening to a wide range of people during my research. Each of the topics summarized in the table below, and no doubt many more, seemed to at least some residents of California I encountered to be important to stem cell governance during CIRM's ten-year mandate. Strikingly, despite the framing of stem cell research in Proposition 71 as a pro-science, Silicon Valley-type investment in real estate, research, and expertise that would provide profitable cures for the people and a massive trickle-down boost to the economy, this did not capture the range of concerns I discovered among those more and less closely associated with the research. Again and again, it was brought home to me how big the gap is between people's emotional and intellectual and social commitments at different times in different circumstances with different information, and the way an issue has to be packaged to get through the political process. In stem cell research, as elsewhere, there is something profoundly undemocratic in the political process, losing as it does the nuances and complexity of people's views. This need not mean, however, that the plurality and complexity of people's experiences and opinions is unable to affect elements of the politics of science as they unfold. The other striking aspect of the curriculum collecting process was how generalizable the end result was: some of the concerns were specific to stem cell research and California in the early twenty-first century, but some were relevant to things far beyond stem cell research.

I believed—and still do believe—that many of the issues raised in this curriculum summary were urgent and were chronically under-debated. It quickly became clear, however, that this curriculum could not be taught in the few hours allotted to the CIRM scholars' ethics and law class as part of our training grant, and that in any case this was probably the wrong group to whom to teach these topics. The science and clinical fellows (from our clinical partner, the Children's Hospital Oakland Research Institute), bright and engaged as they still valiantly managed to be when they made it to class, were faced with a requirement that was, from their point of view, just one more requirement on top of their own heavy research loads. They might have found a course in research integrity or one in intellectual property for the post-Bayh-Dole generation of CEO-scientists in training more germane; in any case (and I am smiling as I write this, thinking of my naiveté when I first imagined such a course without so much as a thought for the circumstances of those who would be asked to take it) they probably did not quite feel that at their stage they were personally responsible for fixing the broken health-care system or overcoming recalcitrant health disparities. While teaching the class, I quickly realized that I was having tiny bits of what should have been longer conversations with a dozen exhausted graduate and post-doctoral students trying desperately to get results within an academic year or two of funding, in mostly new labs. Still, we had some excellent discussions and debates.

TABLE 3.3
Sample of "ELSPETH" stem cell curriculum.

	Course topics (based on responses and concerns collected from science and social science colleagues, students, activists, members of the public)
Ethical issues	Life: When does life begin? Can one life ever be sacrificed for another life?
	Cures: How can progress toward cures be sped up (faced with legal, political, scientific, economic barriers)?
	Justice: How can the fruits (therapies, information, knowledge, jobs, real estate, profit) and risks (of donation, clinical trials, experimental therapies; of investment of time and expertise and money) of science be shared equitably among those paying for the research (the people of California)?
	Humanism: Should there be any limits to science?
	Individual and group autonomy: How should informed consent be obtained, when disease outlooks can be desperate, expectations for therapies can be unrealistic, and emotional and economic pulls (e.g. to be a donor) can be strong? How about those not in a position to consent on their own behalf, and how about when a group's or community's wellbeing might be at stake?
	Gender: Which women, and with what compensation, and with the help of which medical protocols, could provide the eggs for somatic cell nuclear transfer?
	Protection against rogue uses: If induced pluripotent stem cell research goes "garage" (so that anyone can take anyone else's skin cells off any surface where they've been shed, and reprogram them by adding the right transcription factors in their own "garage"), how can each of us make sure no one is cloning or patenting or weaponizing us?

TABLE 3.3 *(continued)*

	Course topics (based on responses and concerns collected from science and social science colleagues, students, activists, members of the public)
Legal issues	Human rights: Is there a right to health care, do rights to health care apply to therapies not yet available, and is there a human rights rationale for funding stem cell research? Do underserved communities have rights to be free of health disparities?

Patenting: What can be / should be able to be patented: body parts, tissue, genes, novel processes for endowing cells with capacities found in nature (like hESC and iPSC techniques)? How should incentives to develop therapies be offset by the need to make therapies available to all at home and abroad? Given the pro-cures mandate, how should potential patent thickets or failures to capitalize on promising patented discoveries that slow the path from bench to bedside be handled?

Intellectual property: To whom should the intellectual property belong from CIRM funded research? To the researcher, to the researcher's institution (whether public or private?), to CIRM, to the State of California? Is the 1980 Bayh-Dole Act, the University and Small Business Patent Procedures Act, which allows US universities to have intellectual property rights in research that was wholly or partly federally funded, the right model at the state level?

Reproductive law: What are the relations between human pluripotent stem cell research and abortion, infertility, and fetal tissue laws? Will the spotlight on human embryos generated by the funding for and media exposure of stem cell research diminish reproductive privacy and a woman's right to choose, by, for example, giving heft to embryo personhood initiatives? Will the clandestine nature of science's traffic in fetal tissue push some parts of stem cell research undercover too?

Cross-border law: How are cells, donors, patients, scientists, knowledge, law, and infrastructure to be monitored for exploitation, equity, freedom of information, regulation, and security, as they move across borders between US states and among nation states? How should conflicts between Federal and State law and between different nation states be reconciled?

	Course topics (based on responses and concerns collected from science and social science colleagues, students, activists, members of the public)
Social issues	Health disparities: [How] should funding for and the pursuit of stem cell research tackle the socioeconomic, ethno-racial, regional, and other patterns of health disparities? Will stem cell research alleviate or make worse the health care options of those currently medically underserved?
	Public understanding of science and scientists' understanding of the public: How do different individuals and groups understand medical conditions and scientific research? What kinds of behavior and protocols promote trust and communication between the public and scientists and clinicians? What models of input and representation best diminish deficits of understanding in both directions?
	Biosociality, biomedicalization, and biocitizenship: What new kinds of sociality are arising in this era of disease activism? Are there new social hierarchies developing between medical conditions as advocates compete for representation and scarce resources? What are the risks and benefits of accessing state resources through one's biomedicalized body?
	Disability justice: How should the different voices of disability rights activists be sought and attended to? How can limits to the idea of cures and a comprehensive approach to disability justice be combined with strong support for science and the search for cures?
	Social contract: What is the new social contract between society and science enacted by Proposition 71? What obligations does science have to (which) publics, and vice versa, under its terms?
	Media: How should the media and campaigns responsibly cover news and views of stem cell research and its publics, without raising unrealistic expectations or falling into over-determined framings?

TABLE 3.3 *(continued)*

	Course topics (based on responses and concerns collected from science and social science colleagues, students, activists, members of the public)
Political issues	Ethical filibustering: Could the CIRM effectively be shut down by legal challenges? Democracy: Is the California Proposition process profoundly democratic, because it requires signatures from ordinary people, or profoundly undemocratic because it takes a lot of money to mount a proposition and because it is prohibitively difficult to repeal one once passed, if its effects are undesirable? California state exceptionalism: In what ways is a large, economically significant state like California able to negotiate with other nation states and the federal government almost as if it were a nation-state in its own right? Does California's pro-innovation climate make it a different (more positive) environment for social and scientific experiments like Proposition 71 than other states? Good governance: How should conflicts of interest standards in the Independent Citizens' Oversight Committee (ICOC) and in the Working Groups and in Grantee institutions be defined and enforced? Should the Bagley-Keene Open Meeting Act apply to Working Group meetings? Can scientific peer review, proprietary research, and open meetings (the gold standards of science, R&D, and democratic politics, respectively) co-exist, and if so, how? Federal / state relations: How does federal and state legislation on stem cell research differ and how should conflicting legislation be adjudicated? Is statewide direct democracy (bond issue) the right way to decide science funding priorities? Regulatory frames: How do other states and other countries organize and regulate stem cell research and what can we teach them and learn from them?

	Course topics (based on responses and concerns collected from science and social science colleagues, students, activists, members of the public)
Economic issues	Allocation of scarce resources: Is stem cell research a good investment for the people of California's overall health? Research into which diseases, and with what rationale (for example, the lowest hanging fruit or most scientifically promising, or the greatest potential profit to the state, or the greatest potential to alleviate suffering, or the greatest potential to improve the health of the greatest number of people), should be funded? Beneficiaries: How and to whom or what should royalty streams from inventions funded with Proposition 71 money be directed? How can therapies be made affordable to all? Fiscal responsibility: How will California taxpayers recoup their investment in the research? Of the various kinds of moneymaking potential of Proposition 71, will any of them pay off, and to whom, and how, during or after the ten year life of CIRM? Will the tax base grow through Proposition 71 related economic activity? Will long term health care costs decrease if successful stem cell therapies are developed? Will there be patent income? Ethical arbitrage: How can federally funded and CIRM funded equipment and research be kept separate without exorbitant outlay, if some research can be funded by CIRM funds but not federal monies? Capital advantage: Is it possible to insure that less well endowed institutions that might be unable to sustain capital investments after the 10-year period of CIRM funding, or that might not be able to offer cost-sharing, will not be discriminated against in awarding grants? Efficacy: How should grant disbursement be conducted to ensure both maximum scientific and therapeutic pay off and maximize repayment to the General Fund?

TABLE 3.3 *(continued)*

	Course topics (based on responses and concerns collected from science and social science colleagues, students, activists, members of the public)
Theological / religious issues	Ecumenism: Is it possible to foster the understanding of different religious views and adherents' bases for different bioethical positions around stem cell research, or must all opposition be taken as anti-science? Can different kinds of moral and religious leadership on stem cell research help guide practice and policy, or is it only possible to include at the table representatives of those religions that are pro-embryonic stem cell research?
	Tolerance and pluralism: Is it possible to have a tracking system in place for treatments from human cell lines that informs all downstream users about their provenance and procurement, so that individuals can access treatment without compromising their religious beliefs?
	Life: What do different religious perspectives say about the beginnings, ends, and quality of life, and how does that bear on stem cell research?
	Cures: There is a deep call in many religions to seek cures and to help alleviate the suffering of others.
	Alternatives to cures: Religious leaders often provide a voice for alternate therapies, and frequently provide all too rare ways of thinking about disease and injury in different idioms and where human bodily limitations can be accepted.

	Course topics (based on responses and concerns collected from science and social science colleagues, students, activists, members of the public)
Historical and geographical issues	Knowing our history: We need to understand where our science and values have come from to appreciate where they are now and where they are going in the future. What kinds of things might we have got right and what kinds of things need changing, and how does our history help or hinder that?
	Comparative perspectives: How do our beliefs and practices differ from the beliefs and practices and knowledge of others around the nation and the world?
	Learning from the past: How do/should historical events and their legacies, such as the slave trade, colonialism, the Holocaust, and Tuskegee, continue to inform our bioethical and scientific views and practices? How can we learn from them?
	The historical role of science in the nation and the state: What role has science played in US and California history, and what is different and what more of the same in Proposition 71?
	Documentation: It is imperative that we document this history-in-the-making represented by the passage of Proposition 71.

In another effort begun in 2005, I and the Science, Technology, and Society Center at UC Berkeley (of which I was a co-director at the time) joined with the Greenlining Institute, a Berkeley-based non-profit dedicated to full economic participation for ethnic minority and other underserved communities, to put together a conference that we called Toward Fair Cures. (See chapter 1.) We invited community and academic experts in diversity in science education, minority-run businesses, women's rights, the history of race and medicine, and disability rights. Dr. Bert Lubin, the charismatic director of the Children's Hospital Oakland Research Institute, agreed to host the conference in CHORI's historic building, formerly a school and a Black Panther medical center.[30] This conference, along with a second conference we co-hosted with UC San Francisco, was my second effort to alter the terms of debate.

OF WOMEN?

My third foray into intervention during the aftermath of the passage of Proposition 71 took the form of a thought experiment. In January 2005 I received a phone call in which I was informed that I had been nominated to a committee, the Human Stem Cell Research (HSCR) Advisory Committee, which came into being as a result of Senate Bill 1260, which had first made hESC legal in California (Chapter 506, Statutes of 2003; Health and Safety Code Section 125118.5), and was established in 2005. The Department of Health Services (DHS) physician who called me told me that my name had been put forward as someone who would bring to the committee expertise in issues in human stem cell research that affect women. I would, in effect, be representing the interests of women, whatever that might mean. Insofar as CIRM would have its own standards-setting committees and procedures, the job of the HSCR Advisory Committee vis-à-vis CIRM was unclear; some thought it had been rendered obsolete or redundant and would be left to "sunset" with the senate bill itself in January 2007. Others felt that the new stem cell initiative could do with some independent oversight and that this committee might provide that, and that California needed a separate committee to ensure standardization and harmonization among all human embryonic stem cell research in the state, regardless of its funding source and beyond the time line of CIRM. The HSCR Advisory Committee went on to develop statewide guidelines for human stem cell research that the California Department of Public Health finalized in 2007, and was asked to continue its work after the 2007 sunset date.

From what I subsequently discovered, it appears that I was nominated as a replacement for a women's-health activist who had sat on the earlier cloning committee, on the grounds that I was more positively disposed than she to stem cell research and the stem cell initiative (although I had not signed up to support the proposition, primarily out of concerns about state funding priorities and health disparities, but also out of concern about overusing gonadotropins to generate leftover embryos) and thus less controversial than she and other possible candidates representing women's-health activism.[31] As it turned out, after a shake up at the Department of Health Services, they decided that there was no need for a women's-health advocate, so my nomination (like hers) was moot. The new committee finally met on February 24, 2006, just before the hearing of the next round of litigation by pro-life groups against CIRM on Monday, February 27, and as Senator Deborah Ortiz was in the midst of submitting a new bill amending the terms of Proposition 71. The sitting committee, predominantly elderly, white, and male, received a knowledgeable tutorial on federal versus CIRM stem cell guidelines from University of Wisconsin-based stem cell research lawyer and consultant to CIRM, Alta Charo, resulting in the situation that the agency supposedly being guided by the state-wide HSCR committee's recommendations was teaching committee members the basics of the regulatory situation.[32]

At that meeting, I began a thought experiment that became one of the guides in my conversations in and around stem cell research: What if one were to take my original charge for the advisory committee seriously, and ask what it might mean to pay attention to the needs of a group called "women," which doesn't present as a group at all, let alone a unified interest group, as regards stem cell research? How could one represent (or whatever the right relationship might be) women's views and lives in regard to stem cell research statewide?[33] I considered issues in stem cell research that emerged as "women's issues" and asked which women they affect and how they might best be navigated. I also considered issues of major significance to some women but which dominant framings of the stem cell debate leave out of the picture. I pursued the following questions (by no means an exhaustive list of women's issues, but resonant with the emerging curricula goals listed above) in relation to class and race as well as to gender:

Are women represented among stem cell researchers and biotech innovators?
Why aren't abortion politics a *women's* issue in the stem cell debate?
Why has egg donation arisen as the premier women's issue in stem cell research when it has had a very different valence in reproductive technologies?
Did authorizing the expenditure of an additional $3 billion, or approximately $6 billion after interest, of state funds on stem cell research adequately represent women's health-care priorities?
What are the special women's roles, particularly as patients, care givers, and mothers, in patient activism and disability rights and why are they heard less than would be expected?
What are the women's issues at stake in women's opposition to, as well as support for, stem cell research?

One might have expected that a major concern about women and stem cell research would have been to make sure that women would receive their fair share of state-funded stem cell training grants and biotechnology research and development. In fact, though, in California after the passage of Proposition 71, this issue was hardly raised. Why women did not organize around this issue is not entirely clear to me, but one reason may be that the life sciences, if not the biotech industry, already attracted a high number of women scientists relative to other science disciplines, and that prominent women scientists were already in the field, even if they were not the national spokespeople for stem cell science or the major beneficiaries of stem cell industry.[34] Scholars and activists concerned with women in science, then, for the most part remained quiet.[35] Women scientists serving on stem cell committees tended not to identify themselves as experts with a particular interest in supporting women in stem cell science or as committee members with special insight into women's issues around stem cell research in general. Perhaps their reticence derived from a concern to be taken primarily as scientists rather than as women scientists (a reasonable concern), and it does not seem fair, in any case, to load yet more requirements on women scientists. Before and after our Toward Fair Cures conference, we met with members of the public who were concerned that Proposition 71 money for training scientists wouldn't reach lower-income communities, and concerned about racial and class disparities in grants allocations. An intersectional analysis including gender was mainly evident around the question of egg donation and not around the question of women in science.[36]

Abortion politics lurks behind debates about human stem cell science in many countries, often dictating its legality and conditions of work, restricting public and private R&D funding and future markets, and pitting religion against science. In the United States, as has already been noted, the opposition to abortion from evangelical Protestantism, even more than Catholicism, is politically powerful. The United States is typical in that religious opposition to abortion extends to human embryonic stem cell research.[37] Since the passage of Proposition 71 promised public funds for embryonic stem cell derivation that would destroy embryos in the process, abortion politics was central from the start. And yet, abortion has not at any point been a women's issue in California around the field of stem cell research.

Paradoxically, even though women have abortions, and even though protecting the right to a legal and safe abortion is known as "a woman's right to choose" and has been the signature political issue for pro-choice women for over a quarter of a century, the connection between embryonic stem cell research and abortion didn't have much to do with gender or women. Women who oppose abortion also commonly oppose stem cell research because it destroys embryos, not because of women's reproductive health or women's autonomy. The political movement to bring leftover embryos from infertility treatments to term invoked right-to-life women as the literal and symbolic mothers/wombs willing and capable of preventing the "genocide" represented by destroying the unused embryos, bringing anti-abortion back to gender and family politics. Embryonic stem cell lines do not exist until after the embryo entity is sacrificed, however. This means that once stem cells are in question there is no longer any embryo that a right-to-life mother could bring to term. Anti-abortion conviction, then, is still the major impediment to support for embryonic stem cell research and yet it is not really a "women's issue," even for right-to-life women, except indirectly.

Many women who support stem cell research also support legal abortion, and vice versa, but it does not appear to follow logically, at least in California, that these are two parts of the same moral framing in the way that those who oppose abortion also tend to oppose stem cell research. There were, as far as I could tell, three main reasons why not all pro-choice women were supporters of embryonic stem cell research and/or Proposition 71. The main reason appeared to be that many women, especially feminists, were major participants in articulating opposition to reproductive and genetic technologies on different grounds from those of right-to-lifers. Some pro-choice women worried about the special risks to women posed by egg donation for embryonic stem cell research. A second reason was that a major appeal of stem cell research was its promise of cures for disability and disease. Though many find this reprehensible or at least fraught, the right to abortion has gained part of its legitimacy from cases of "therapeutic abortion" for disability; offering pregnant women fetal testing for conditions like Down Syndrome is predicated on the availability of abortion following a diagnosis of the difference. New conversations about interventions to alleviate disability through stem cell research threatened to bring to light this relatively unexamined part of the U.S. abortion rationale.[38] A third reason why some women's pro-choice views did not always translate into pro-stem-cell views came from a desire to protect reproductive privacy. Many scholars and activists have fought for a gradual extension of reproductive and sexual privacy protections over the years, allowing such categories as single women, gay and lesbian lovers and parents, mixed-race couples, and people with disabilities to reproduce and have access to reproductive technologies free from state intervention and discrimination in their private lives.[39] The intensely public issues involved in stem cell research regulation risked undermining these hard-won trends in reproductive privacy.[40] While support for abortion and support for carrying out stem cell research often go together, then, some pro-choice women considered embryonic stem cell research and abortion to occupy different moral realms.

Thus, despite the fundamental role of abortion politics in stem cell research and the gendered nature of abortion itself, "a woman's right to choose" did not effortlessly encompass support for stem cell research, and the intertwining of stem cell research and abortion did not, *prima facie*, present as a "women's issue" on either side of the abortion debate. The question remained as to whether abortion politics in the stem cell debate *should* have been a women's issue. If one were representing women in the stem cell debate, should one have picked up and argued the threads that in fact emerged as women's issues, or was it at least as important to try to figure out what differently situated women ought to have a say in? My fieldwork suggests the latter. If someone had been given the opportunity to try this, it might have involved the following: articulating anti-hESC, pro-life women's concern to distinguish among stem cell lines as research goes forward, so that they can access the fruits of research, in the form of products or testing systems, and eventually cures for themselves and their families, that did not involve sacrificing embryos. A corollary to this would include making the argument that allowing pro-life citizens access to stem cell therapies that did not destroy embryos could lead to good labeling, tracking, and accounting practice for human biologicals in general as they increasingly enter product streams. This kind of tracking infrastructure might, in turn, facilitate international safety and standardization and end up improving stem cell science. Again, a general truth here is that opposition can often be translated into standards of care that *improve* rather than *impede* research, as long as the specifics of opposition are internalized into technical procedures rather than externalized as political barriers to research.

Pro-choice women's concerns about women's experience could also be insisted on. My fieldwork suggests that a mother who has undergone pre-implantation genetic diagnosis and has leftover embryos that have or are carriers for a fatal disease might well be a compelling witness to the value of those embryos for research and the problems associated with attempting to bring them to term. Pro-choice women articulate both the differences between the right to reproductive privacy including safe legal abortion and the right to conduct stem cell research, making sure that the unique role of a woman's body in the two cases be kept in mind. And again, one could drive policy and practice in ways that are designed to mitigate rather than exacerbate other concerns that some pro-choice women have expressed toward stem cell research, such as resistance to eugenics and cloning, and concerns about the health disparity implications of the research. This, too, would probably lead to better, not worse, research and care. Women in favor of hESC research could learn from pro-life opposition to the research about dangers for disability rights implicit in prenatal diagnostic procedures and therapeutic abortion, and those against the research on anti-abortion grounds could learn from women advocating for stem cell research about the latter's trials and suffering that prenatal diagnosis and stem cell research has the potential to alleviate.[41] The different roles that gender could play in the link between stem cell and abortion politics should be studied and appropriated to inform policy and practice.

For a range of interesting reasons, protection of potential egg donors became a signature ethical concern in the politics of stem cell research in California and elsewhere, and emerged as *the* women's issue in pluripotent stem cell research ethics. Furthermore, it was one of the few ethical issues that garnered bipartisan support. Fueled by the South Korean stem cell scandal where not only were research results fabricated but women were coerced to donate eggs, it also functioned as a chip in a game of national and individual rivalries over supremacy in stem cell research (explored in the next chapter).[42] Yet protection of egg donors was not *ab initio* a universal ethical standard in barely emerging national and transnational regulatory and trade regimes for human stem cell research. In California, as elsewhere, women's-health advocates played a role in bringing this issue to the fore.[43] On July 1, 2005, the Pro-Choice Alliance for Responsible Research and the Center for Genetics and Society, both California-based civil-society groups favoring human embryonic stem cell research but advocating for effective regulation, wrote a joint letter to the members of the Scientific and Medical Accountability Standards Working Group at CIRM in which they recalled "the unacceptable price paid by racial and ethnic minorities, women, the disabled, and other vulnerable groups in past scientific endeavors" and the National Academies' acknowledgment of "the serious risks of the hormones used in conjunction with surgical egg extraction" in its stem cell research guidelines.[44] They noted that "women who provide eggs for stem cell research will effectively be the first research subjects under the CIRM program," and recommended that "volunteers and recruits who provide eggs for research should receive no-cost medical treatment for any condition arising out of the egg extraction process."[45] Hands Off Our Ovaries (an international organization formed specifically to oppose natural resource extraction for scientific research from the ovaries of young women) saw California's stem cell initiative, in the context of the Hwang Woo Suk egg scandal, as a primary target of their advocacy.[46]

Despite how over-determined this concern for women egg donors came to seem in the years immediately after the passage of Proposition 71, it did not simply follow from the state of affairs in egg donation as it was already being routinely practiced for assisted reproductive technologies. By the time of Proposition 71, oocytes had been removed from the bodies of women for their own and others' reproduction for a quarter of a century in the United States with very little outcry about donor protection. Although egg donation in reproductive technologies had been in the news regularly, egg-donor coercion or risk bearing had not been a major aspect of comment or censure in the mainstream press.[47] Usually, it was the sky-high prices for eggs from certain kinds of donors that had been the subject of debate; occasionally, it was the multiple pregnancies and/or the stockpiled "leftover" embryos that resulted from ovarian-stimulation regimes tailored to produce large numbers of eggs at once.[48] From the early days of reproductive technologies, feminist and anti-racist voices had been raised about the potential for exploitation in the fertility industry along class, race, and national lines with both egg donation and gestational surrogacy.[49] From time to time a woman suffering severe ovarian hyper-stimulation had hit the news, but it usually formed part of a more general condemnation of trafficking in women's bodies and was not part of a call to protect egg donors.

In the United States, the Ethics Committee of the American Society for Reproductive Medicine, working largely out of the public eye, issued guidelines in 2000 on acceptable payment to egg donors that, it asserted, did not represent "undue inducement" to donate. The Ethics Committee's principle concern in protecting egg donors was to discourage the trumping of rationality as regards the risks of the procedure by the prospect of excessive financial gain.[50] This could be averted, members of the committee suggested, by capping how much donors should be paid. Members of the Ethics Committee also expressed concern about eugenics and the commercialization of childbearing, but again their main interest was to keep costs down to what might be understood as reasonable reimbursement for the time and costs of undergoing hormonal stimulation and surgery for egg extraction.[51] Above all, they strove to present a medical sub-specialty governed by good motives—both doctors and donors helping those less fortunate to have babies—and run within the bounds of rationality, informed consent, and affordable private treatment. According to the Ethics Committee's report of 2007, "recent scientific developments suggest that oocyte donation may become an important process in the field of stem cell research." The principle that "recruitment practices incorporating remuneration sufficiently protect the interests of oocyte donors," argued for in the committee's earlier report, was reiterated, but it was interpreted differently. Instead of worrying about the rational capacity to gauge risks and benefits being outweighed by financial incentives or access to fertility treatments, the committee argued that egg donors had a right to a financial incentive, just as sperm donors and participants in clinical trials had, as a matter of fairness, and that financial reward for time and trouble is entirely compatible with donating eggs for the right reasons.[52] Their primary concern had switched to being one of ensuring an ethical supply in the face of rising demand.[53] In short, even though concerns about "eggsploitation" of donors coincide on the surface, the debates around egg donation for reproductive technology have tended to be different from those around egg donation for stem cell research.[54]

The differences stemmed from the public nature of stem cell research versus the private nature of assisted reproductive technologies (ARTs), the fact that in ARTs the eggs stay within the context of reproductive intent whereas they exit it for stem cell research, and the stratified nature of egg donation for ARTs. Needing the popular vote, being funded with taxpayer money, and being under the strict ethical and licensing regimes of scientific research all made egg donation for stem cell research subject to much more public scrutiny and oversight than ARTs, which had always managed in the United States to work largely out of the public eye. Drug companies, too, had benefited from the relative lack of public scrutiny of reproductive technologies. Whereas many corporations did not dare to invest in human embryonic stem cell research, drug companies had found fertility drugs to be very profitable, and innovation—for example, recombinant versions of the drugs used for ovarian hyperstimulation—had been lucrative.

Many body parts and body products can be gifted or sold apart from the body of which they were once part (for example, organs, hair, blood, sperm, eggs, milk); many can be sold, although not always by the donor (for example, blood, sperm, eggs, some organs); many are used for medical diagnostic procedures (for example, urine, feces, blood, embryos); many are used in industry to generate value-added products for forensics, security, wig making, medical or fertility treatment, and so on (for example, teeth, blood, fingernails, skin, hair, urine, eggs); milk is food; many are rarely and only with difficulty designated as waste (for example, cadavers, organs, eggs, embryos) while others are often waste (for example, blood, sperm) and others are usually waste, and their management as such requires an enormous infrastructure worldwide (for example, urine, feces, hair); several involve considerable short-term or long-term risk to the donor (for example, living-donor organ donation, skin, eggs) while many normally do not involve much risk (for example, post-mortem organ donation, hair, blood, sperm, urine, fingernails); many are sometimes sacred and sometimes profane (for example, organs in a dead body, which are usually sacred and their taking would be a desecration, but which can be extracted under the right consent and medical conditions; hair, which is sacred to members of some religions and profane to others; embryos); some can only be donated by certain members of the population and not others (for example, matched tissue; sperm, eggs). Each part of the body has a different history and pragmatics of alienability that itself varies from one part of the world to another.[55]

Since they have been circulating outside the body in significant numbers, women's eggs have displayed characteristics that are shared, but in different combinations, by different body parts. The donation or sale of body parts and products always faces, but to varying degrees, questions about the voluntarism or coercion of the donation or sale; questions about the beneficiaries of the value-added economy of the body part; the risk of biological, social, or transnational segmentation and stratification of the donor and recipient pools; the urgency or moral appeal of the donation or sale; and whether the body part donation or sale is part of a preexisting bureaucracy and whether that bureaucracy is seen as having legitimacy. These things are not unique to egg donation but they have surfaced in the stem cell debate in a particularly public manner generally uncharacteristic of other body parts. Using women's eggs for stem cell research means taking the eggs out of their reproductive context, which also brings into question the narrative of the gift of motherhood that commonly accompanies the act of egg donation for fertility purposes. It also moves the eggs out of the private and intimate domain of the clinic, into the world of publicly funded research. Not surprisingly, procuring eggs to do research conducted with taxpayer money, struck many as needing a new kind of scrutiny and a particular concern that the state not be the agent of gendered bodily harm. For the civil-society groups and women's-health groups mentioned above, drawing attention to egg donation for state-funded research was a way to begin the process of regulating egg donation in general, including in the fertility industry.[56]

Proposition 71 banned payment to egg donors, and Senator Ortiz sponsored a bill that further restricted compensation to women who donate eggs for stem cell research to ensure that coercion is not involved. What were the arguments behind this decision? One line of thinking has it that if donors are compensated, it amounts to paying for eggs, which is perilously close to paying for babies. There is overwhelming consensus that paying for babies (the goal for eggs in reproductive technologies is to make babies, but not in stem cell research, unless the embryo from which the stem cells are extracted is seen as a baby) should be prohibited because it amounts to trafficking in human beings, and trafficking in human beings is slavery.[57] Paying for eggs seems also to mean that eggs are commodified. The anti-commodification argument, as long as it is not spelled out too clearly, has broad support across the political spectrum. This is one reason for surprising unity around egg-donor protection that emerged after the passage of Proposition 71. Three of the objections to treating eggs as a commodity are the following:

• that eggs and embryos are sufficiently people-like, or inextricably connected to and thus not completely alienable from people, and people are never commodities; thus, treating humans like commodities implies treating them as instrumental, rather than as means unto themselves

• that commodifying eggs introduces market dynamics that can make the world's most vulnerable subject to coercion by rent-seeking parties and encourage trafficking in eggs

• that there is an ethos of a gift economy in tissue and organ donation that aims to provide cures, such as stem cell research, and to commodify eggs would be to subvert the emotions and motivations of patients and donors alike.[58]

Is all payment commodification? There are many examples legally recognized as such and informally honored as such where it would be a stretch at best to say that paying for something is the same as commodifying it in the sense of the critiques above. For example, if Ms. X pays her mother to take care of her toddler at home three days a week, most people would probably say she is respecting her mother's labor by paying her, not commodifying her mother's relationship with her grandchild. Similarly, there are many examples of things that are not entirely commodified despite being paid for, such as the use of people's names, or licensing techniques that are patented; license fees are paid for use, but the innovation or the individual-referring name does not leave its original owner. Modeling egg donation on either of these kinds of things would work relatively well in averting the critiques of commodification. The other option is resolutely to enforce that it is effort or expense of donation that is compensated, and not the eggs themselves.

Refusing payment for egg donation has a consequence that it prohibits the donor from acquiring any profit derived from their eggs, while those who commercialize cell lines derived from the eggs are free to profit.[59] If companies and even individual researchers at public universities can profit from state investment in stem cell research, is it fair that egg donors themselves cannot? To be concerned about commodifying eggs because that would rob women of their essential human dignity, one has to believe that body parts should be treated in significant ways as persons. Those on the left who worry about commodification tend to get some of the moral heft of their concerns from these meta-ethical considerations, but have more practical problems in mind. They may be worried on the one hand that the commodification of body parts signals the end of our human future, threatening a fundamental condition of humanity, the valuing of people in and of themselves, not because of their instrumental or other use value.

On the other hand (and with very good cause, in view of some organ-donation markets), anxiety about commodification is part and parcel of anxiety about coercion. From Indira Gandhi's state-sponsored vasectomy campaign in the 1970s to accounts of the kidnapping of street children for organ donation in Brazil and elsewhere, trafficking in human tissue is a real danger when there is a market for body parts. If there is a market for eggs, or for any other human tissue, unscrupulous third parties will go to any lengths to attain those eggs. This kind of traffic is extremely hard to police because it preys on the most vulnerable people in society and tends to be part of an informal or illegal economy. The economic calculus is simple: when a person is worth little or nothing in society, any value attached to his or her organs or body tissues is value added. Harvesting eggs, if the eggs themselves (rather than the act of donation) are paid for, makes economic sense even if, in the most desperate of circumstances, it sacrifices the donor.

Although the protection of egg donors is extremely important (I have advocated for strict protections, particularly around potential long-term effects on donors of ovarian hyperstimulation, around repeat donation, and around the childbearing status of donors) and is an issue of great importance to women as a class, we should be wary of taking the current framing as the only way to view the topic. In view of how different other donations and sales are, and how different egg donation is in reproductive technologies, it is important to be able to ask other kinds of questions that might give answers that are just as important for the well-being of women. For example, it is important to be able to ask "Which kinds of tissue or organ donation for which purpose is oocyte donation like?" and "What lessons can be learned from donor protection in each of these?" When is something waste, a gift, a commodity, a research substrate, or a combination of these, and how can we move beyond the gift/commodity dichotomy for the research situation? What are bioethical regimes in other countries, and what can be learned from them? Is the "ethics of egg donation" at risk of being a way of imposing a particular bioethical regime on other countries that gives countries with regulation based on protection of human subjects an unfair head start, and perhaps set up transnational dynamics that will themselves lead to a transnational and inequitable egg trade? Which groups can or do donate, with which motivations? Are barriers to donation—for example, being in the right information networks to be aware of the option of donating, or having expenses covered—substantial? How much do research imperatives require demographic variety in the donor population, and what recruitment practices will address this? In particular, is it necessary to have representation of all socioeconomic strata and ethno-racial groups among egg donors for non-SCNT stem cell line derivation to give every California citizen the chance find a good tissue match in a stem cell bank? For disease modeling, rather than therapy delivery, how important is it to get eggs from all groups of women in society? Do levels of reimbursement discriminate against one group or another in ways that affect either the research imperatives or the health and safety of donors? Just as it is essential not to over-induce any groups of women to donate, how do you make sure that enough women from all groups participate in egg donation? How can long-term and short-term health and safety concerns best be met for donors undergoing ovarian hyperstimulation and surgery, for technicians working with the eggs, for the eggs themselves, and for recipients of donated eggs? How can psychological expectations of cures be tempered without suppressing support for a hopeful technology, so that undue pressure is not put on family members to engage in altruistic donation? Egg donation, as a women's issue, really could and perhaps should go far beyond the question of whether or not to compensate egg donors.

In some ways, the question of health-care priorities, access, and affordability is the exact opposite of the egg-donation issue. Access and affordability for stem cell therapies themselves have been addressed in the California debate since the passage of Proposition 71, but health-care priorities more generally have hardly been raised at all in public discussion. I asked several women a version of the question "Does authorizing the expenditure of an additional $3 billion, or approximately $6 billion after interest, of state funds on stem cell research adequately represent your own, or women's health-care priorities?" No one answered that it did, even among women who were strongly in favor of stem cell research. A response I heard a number of times was that if it had been up to them to spend $3 billion on health care it would have gone toward insuring the uninsured and on providing primary care and preventive medicine. Many women from many backgrounds see themselves as especially responsible for their own and their families' health care, and their major concern is its availability.

Debating stem cell research from the point of view of its place in the allocation of scarce health-care resources requires a rather different perspective that is not easy to sustain when the question at hand is the implementation of regulations and policies for an already enacted stem cell bill. One woman told me that "stem cell research is like gay marriage" in her congregation—the pastor asked his congregants one Sunday to stand if they thought that prohibiting gay marriage was a serious issue. One woman stood up. He then asked who thought access to basic health care was a serious issue and the whole congregation surged to its feet. That, she said, represented her views on stem cell research—not only is it less important by far than basic health care, but she thought that it would not affect her predominantly middle-income and low-income African American community much, as most members would be unlikely to be the beneficiaries of high-tech interventions such as stem cell therapies, although personally she found the research exciting. On the one hand, there is a readily identified primary health-care burden on women that makes them question whether this is a good use of funds, and a sense among some women that stem cell research doesn't apply to them or their families. On the other hand, whether or not women say they would have spent health-care money this way, many women articulate support for spending on stem cell research because of the promise of respite from their own or family members' often under-treated chronic diseases and debilitating conditions. Caring for the seriously disabled (whether oneself, one's kin, or someone one is paid to care for) is still a heavily gendered domain, so women disproportionately suffer from and disproportionately witness the suffering that some conditions can bring as well as disproportionately endure the sometimes harsh conditions of being a caregiver, even if they are not themselves suffering.[60] Addressing all these concerns and bringing the debate about stem cell research back into a broader discussion of health-care priorities and access and affordability is a major women's issue, and yet it is barely articulated in the current conversation.

Following on from the last point about women's care-giving roles in relation to disability, it is essential to bring to light the role of women as patient activists and as oppositional activists. Women have been extremely influential as patient activists on their own behalf, and for their sick and disabled children and other family members and friends.[61] To forgo consideration and debate of women's roles as patient activists in stem cell research would be to completely ignore the gender dimensions of the very biopolitical and biosocial movements that led to the passage of Proposition 71. The "direct democracy" aspect of a bond issue, and its success through its so-called heartstrings appeal depended deeply on family and care relations and dedicated parental activism as well as on patients acting through their disease groups on their group's behalf.

Another way in which women have been and remain prominent in the stem cell debate and that needs taking into account in its own right concerns their role as spokespeople in opposition to stem cell research on feminist, anti-racist, and disability-activist grounds.[62] Women activists have articulated gateway fears associated with tampering with germ lines, and with chimerical research. They have pointed out the way that stem cell research could play into the hands of older or still existing eugenic platforms, and bring back race science. They have acted for the protection of egg donors, and for responsible and just policies for access and affordability that will prevent some women's bodies being the pharmaceutical chest for the world's wealthiest. Women have also been important in supporting disability rights and pointing to the risk that stem cell research, as it is frequently represented, seeks to find cures for everything, and as such casts everything deemed a deviation from normal as in need of a cure. Spokespeople for disability justice recognize both the importance of finding cures but also the fundamental risks of casting everything as in need of or susceptible to curing, at the possible expense of putting pressure on the world to accommodate a little better to different embodiments, and to value the different realities of those with disabilities. In California, these groups have been exemplary and thorough in providing challenges to the all too ready assumptions with which we assume that scientific progress is good for (all) women, without working explicitly to ask whether it is.

REAL ESTATE?

Under its second president, the Australian IVF pioneer Alan Trounson, CIRM positioned itself to continue to push for funding research in areas where there would still likely be a shortfall of federal funding even after the Bush restrictions were lifted, like somatic cell nuclear transfer cloning. Capitalizing on Alan Trounson's networks, CIRM continued its practice of forging international agreements with stem cell organizations, by signing (in June 2008) an agreement with the Cancer Stem Cell Consortium of

Canada and the state of Victoria and (in October 2008) an agreement with the United Kingdom's Medical Research Council. Less than five years after the passage of Proposition 71, and an era later, CIRM was busy reassessing its own role while collecting comments on and revising its policies to reflect the new draft guidelines for human stem cell research of the U.S. National Institutes of Health. The "Draft National Institutes of Health Guidelines for Human Stem Cell Research"[63] reflected newly inaugurated President Barack Obama's March 9, 2009 Executive Order 13505, "Removing Barriers to Responsible Research Involving Human Stem Cells," reversing former President Bush's August 2001 policy analyzed above.

CIRM responded to the NIH Guidelines of 7/7/2009 by welcoming them, claiming its place in modifying the draft guidelines so that they recognized the importance of "the global aspect of stem cell science," and pledging to continue plugging the hole of un-fundable things at the federal level, especially parthenogenesis and SCNT, and thereby acting as scientific frontier, possibly making it easier for the federal government to make consensus policy later on. CIRM's press release of July 6, 2009, upon the announcement of the NIH Guidelines, contained the following passage and included updated figures representing expenditures to date[64]:

CIRM looks forward to working with the NIH to evaluate the inclusion of parthenogenesis and SCNT lines in the future. As the Governing Board put forward in their recommendations, CIRM views all sources of hESCs—including somatic cell nuclear transfer and parthenogenesis—as potentially important for advancing regenerative medicine. The organization will continue to uphold the highest standards of oversight for hESC derivation and research not currently eligible for federal funding. This commitment will enable NIH to leverage current efforts in its future programs.

As the Obama presidency progressed, CIRM continued to give out research grants, forge international collaborations, and fund the building of research facilities. In February 2011, the Ray and Dagmar Dolby Regeneration Medicine Building opened as the headquarters for the Eli and Edythe Broad Center of Regenerative Medicine and Stem Cell Research at UC San Francisco. The building cost $132 million, and was paid for by a mixture of public and private funds (though named for its private donors), illustrating perfectly CIRM's desire to build facilities that would anchor stem cell research in the state, as well as to fund research from bench to bedside. Initially justified by the potential need to separate federally funded and state-funded research, because of the different rules governing each, the real estate boom of Proposition 71 did not disappear with the advent of iPS cells or with Obama's reversal of Bush's stem cell policy. Real estate and stem cell research were linked in the person of Robert Klein, and continued to co-produce one another in California.[65]

Likewise, CIRM continued to promote patient advocates as its "public" face, showcasing them at its annual World Stem Cell Days. In what may have been a decisive turn, CIRM followed the period immediately after which Geron pulled out of its Phase One hESC clinical trial (see chapter 2) with a leap into a convergence with genomics, inviting the genomics pioneer Craig Venter to speak on the occasion.[66] Around the world, perhaps the most promising results began to appear in personal drug testing, though some potential transplant therapies were in safety trials.[67] With California's budget no better, and the end of the ten-year period in sight, CIRM underwent an assessment by IOM in early 2012 that was again staffed only by supporters.

New science, especially science with ethics, happens through the combining in politically legible packages of language, feelings, knowledge, money, material, status, institutions, and so on. The potential to change the path of science in line with other concerns does not disappear, however. If it were possible to encourage a habit of thought experiments, linked to a curriculum for schools and colleges sourced from the communities in question and engaging many kinds of experts, other concerns could gather momentum. In view of the high stakes of biomedical funding and research, and of the priorities of health care, this could be a matter of life and death. It might also be a way out of the impasse of being stuck with the bundle of interests that made the science possible in the first place.

STEM CELL BRAIN DRAINS

President George W. Bush's 2001 speech announcing restrictions to federal funding for human embryonic stem cell research was greeted by the press and the scientific community as a blow to the United States' international competitiveness and supremacy in science.[1] Fears of a "brain drain"—a trope that would continue to surface around stem cell research in years to come—were stoked as prominent U.S. stem cell researchers threatened to move overseas to more permissive regulatory environments.[2] Despite the importance of the United States to basic science training and research worldwide, a sense began to emerge that the frontiers of regenerative medicine and stem cell research were up for grabs. This fueled nationalisms at home and abroad.

This possibility that leadership in stem cell research might be more open to contestation than usual came from regulatory and ethical differences in carrying out research with gametes and embryos and genetics, giving those countries permitting and funding more kinds of research a perceived competitive advantage. It was not that the field was newly global or that stem cell research was only just becoming international, however. By the time of President Bush's 2001 policy, human embryonic stem cell research was already widely distributed around the world. The "presidential" stem cell lines, research with which was made eligible for U.S. federal funding by Bush's stem cell policy, had been derived in the United States, Australia, South Korea, Israel, India, Singapore, and Sweden. Other countries were also highly competitive, including the United Kingdom, which figured in the brain-drain stories and which had been at the forefront of assisted reproductive technologies (the birthplace of *in vitro* fertilization) and mammalian somatic cell nuclear transfer cloning (Dolly the Sheep), and was emerging as a pioneer of bio-banking. The idea that the brakes had been placed on the research in the United States, then, didn't so much drive the research overseas, as open up a new stem cell geopolitics where not only other countries, but some individual states within the United States, claimed or were feared to have a new competitive edge over the United States as a national sponsor and beneficiary of stem cell research.[3]

Toward the end of the period covered in the book, President Obama's March 2009 executive order (discussed in the last chapter), as it removed Bush's August 2001 federal funding restrictions, summed up this nationalist competitive logic from the U.S. point of view[4]:

Medical miracles do not happen simply by accident. They result from painstaking and costly research—from years of lonely trial and error, much of which never bears fruit—and from a government willing to support that work. From life saving vaccines to pioneering cancer treatments, to the sequencing of the human genome—that is the story of scientific progress in America. When Government fails to make these investments, opportunities are missed. Promising avenues go unexplored. Some of our best scientists leave for other countries that will sponsor their work. And those countries may surge ahead of ours in the advances that transform our lives. . . . We will ensure America's continued global leadership in scientific discoveries and technological breakthroughs.[5]

In between these two rhetorical points, Bush's fueling of an international competition, and Obama's rehearsal of the trope of the brain drain to claim his policy as one necessary to a scientifically competitive nation, stem cell research became seen in the United States and elsewhere as a field where the usual hierarchies among scientific nations might be rearranged, and around which new global biomedical research competitiveness might be built. In this chapter I explore this idea of competitive advantage in the middle years of the first decade of the new century, from California, as a state within the United States with a more permissive research environment than the country as a whole, to the construction in the United States of an "East" around which the brain-drain rhetoric concentrated. As the decade progressed, several countries and international organizations grappled with bioethical legislation and guidelines around embryo and egg procurement and the limits to permissible research. By the last third of the decade, efforts internal to the international stem cell research community to harmonize and standardize legal, ethical, and scientific practice

TABLE 4.1
Pluripotent stem cell research regulations by jurisdiction. The jurisdictions represented are the ones discussed in this chapter.

Jurisdiction	Regulations and dates
United States	Dickey Wicker Amendment, signed by President Clinton in 1995; annual appropriations bill rider since 1997 President George W. Bush Human Stem Cell Policy, August 9, 2001–March 9, 2009 Executive Order 13505, Removing Barriers to Responsible Scientific Research, President Obama, 2009
California	SB 322, Ortiz, 2003 Proposition 71, California Stem Cell Research and Cures Initiative, 2004 Legislation SB 1260, Ortiz, 2006
United Kingdom	Human Fertilisation (Research Purposes) Regulations 2001 "Human Fertilisation and Embryology Act 2008 (Amended HFE Act 1990)
South Korea	Bioethics and Biosafety Act, 2005 Temporary moratorium on hESC research following Hwang scandal Bioethics and Safety Act (Revised) 2008
Singapore	Bioethics Advisory Committee report "Ethical, Legal and Social Issues of Human Stem Cell Research, Reproductive and Therapeutic Cloning," accepted by Government July 2002 BAC report "Donation of Human Eggs for Research," released 2008
European Union	By country for procurement:80 most allow procurement of hES cells from supernumerary IVF embryos; Belgium, the United Kingdom, and Sweden allow creation of embryos for hESC research; Germany and Italy cannot derive their own hESC lines but can import them for research EU Tissue and Cell Directive, Directive 2004/23/EC EU Regulation on Advanced Therapies, Regulation (EC) 1394/2007
United Nations	United Nations Declaration on Human Cloning (non-consensus), 2005 UNESCO Universal Declaration on Bioethics and Human Rights, 2005

had picked up pace. As the decade drew to a close and into the second decade of the twenty-first century, the brain-drain rhetoric around stem cell research did not disappear, as evident in Obama's speech, but it began to be displaced by a new kind of boundary drawing around rogue versus ethically and scientifically accredited science. Stem cell tourism, in particular, seen as preying on the desperation of patients waiting for cures, was given a distinct geopolitics. Scandals of for profit stem cell clinics pedaling untested therapies with unethically procured human tissue emerged within the United States as well as in and with other countries, but stem cell tourism was largely portrayed in the U.S. as an off-shore hazard to good science.

After President Bush's stem cell policy came into effect, several individual states launched their own stem cell support initiatives that were more permissive than federal policy. Senate Bill 253 was passed in California in 2002. It allowed the donation of eggs and embryos to stem cell research from fertility clinics, introduced a mandate to inform fertility patients that they could donate leftover embryos to stem cell research, and explicitly authorized the creation and destruction of human embryos for stem cell research.[6] Governor Gray Davis announced the bill's signing with a promise of a future state funding call for stem cell research.[7] Being the first state to act on regulatory arbitrage between jurisdictions, California was widely reported as offering its own brain gain destination by passing this bill.[8] Connecticut, Florida, Illinois, Maryland, Massachusetts, New Jersey, Wisconsin, and other states also pursued state initiatives to increase support to some kinds of pluripotent stem cell research between 2002 and 2009.[9] None of these state initiatives was as ambitious as California's second effort, Proposition 71, the California Research and Cures Act of 2004 (discussed in the previous chapter), which allocated $3 billion to stem cell research over ten years.[10]

At the same time as the nation's sense of international competitiveness under threat was driving the devolution of stem cell funding from the federal to the state level, the imagined geography of the competition overseas was coming into focus. Scientists and institutions in several countries in Europe and Asia, in Canada, in Australia, and elsewhere had been major contributors to the field since its inception. In the press and in political and editorial rhetoric, however, U.S. fears about regulatory advantages in stem cell competitiveness began to focus increasingly on Asia around the middle of the decade. The Korean stem cell scientist Hwang Woo Suk published two famous papers in *Science* in 2004 and 2005 claiming to have succeeded in producing stem cell lines from somatic cell nuclear transfer embryos, and then to have made stem cell lines with eleven patients' own DNA. These papers held out the promise of patient-specific cellular repair kits.[11] These claims constituted the kind of field changing breakthrough that struck scientists and the media alike as exactly the kind of achievement that could have happened in the United States had funding not been restricted. On May 23, 2005, the week of the publication of the second Hwang *Science* paper (a week I remember very well), Robert Klein, the Chairman of the Board of the California Institute of Regenerative Medicine, called it "terribly frustrating" that the CIRM bond was still mired in legal challenges while the Koreans were forging ahead, and Zach Hall, then CIRM's interim president, said:

What the Koreans are doing makes it clear that this work is going to go on and it's going to go on all over the world. . . . We're poised and ready to go. We could be doing that here. We need to be doing that here . . . and we're not even out of the starting blocks.[12]

Why was it that Hwang had apparently got there first? What was it about him and his colleagues and their scientific working conditions that accounted for the breakthrough being Korean rather than Californian or American? East Asian countries were frequently portrayed as having a more favorable environment for research, a "regulatory oasis," supposedly enjoying a freedom from religious and ethical scruples about research involving human embryos, and blessed with governments eager to support a science economy. This rose to a peak in 2005, and again—though this time coupled with Europe—in 2007.[13]

By the time CIRM was up and running, when U.S. based journalists and scientists talked about the threats to international competitiveness and the risk of brain drains down regulatory gradients, it was increasingly common to hear certain Asian countries added to the United Kingdom and the state initiatives as brain-drain destinations. Stem cell scientists working in Asian labs participated in and responded to the rhetoric, emphasizing not just favorable regulatory environments, but also commitment, work ethic, government support, and the international spread of stem cell research excellence. An article published in *Business Week* in January

2005, "Asia Is Stem Cell Central," perfectly captured the movement of the brain-drain rhetoric to Asia:

Singapore and others are racing to grab the lead in a promising field. . . . Alan Colman . . . chose the city-state because of its tolerant climate for research using embryonic stem cells. . . .The government has established a $600 million fund to invest in startups. . . . Last year, Singapore opened Biopolis. . . . Singapore isn't the only country in the region trying to profit from the U.S. restrictions. . . . The progress the Asians have made is "astonishing." . . . Still, Asian countries are far from assured of leading the way in stem cells over the long term . . . a lax approach to oversight and ethics in some labs, including the use of stem cells drawn from fetuses aborted in the second trimester in China. . . . More worrying for the Asians is the growth in alternative sources of funding for stem cell research in the U.S. . . . California approved Proposition 71. . . . Seoul, for instance, has dished out a total of just $27 million. . . . Singapore and other countries also pales in comparison to what California plans to spend. California may not be the only worry. Britain has a relatively liberal policy toward stem cell research. . . . "I don't think anyone country can monopolize stem cell research," says Susan Lim, chairman of Stem Cell technologies, a Singapore startup. . . . California's research effort will attract attention, but "Korea, Singapore, and China will be even more committed to pursuing it," says Hwang Woo Suk.[14]

After the furor that resulted from the late 2005–early 2006 revelations that Hwang and his colleagues had lied about the eggs used in his lab's research and that the supposed patient-specific stem cell lines were not in fact SCNT lines had died down, and after the *Science* papers had been retracted,[15] Asia continued to be named as threatening California's and the United States' competitiveness if anti-stem-cell lawsuits and stem cell regulations in the U.S. were not overcome. The rise of multidisciplinary regenerative medicine research sites such as Singapore's Biopolis continued to stoke the idea of an Asia ascendant.

An article that appeared at the very end of 2006 on the "stem cell race," portrayed the evolving post-"Hwanggate" geopolitical landscape from a California perspective.[16] With the subheadings "No Clear Leader," "Priceless Support," and "Losing Ground,"Terri Somers of the *San Diego Union-Tribune* captured a multi-national field, with many emerging leaders. Although the article no longer used the language of a brain drain, the writer and the stem cell scientists interviewed remained concerned with what other countries were doing "while the United States is sidetracked by a political and moral debate," and while CIRM monies were held up by legal challenges despite California being home to "50 percent of the world's biotechnology research." The report characterized Asia and Europe as regions staking claims to leadership of the field, but also included countries outside these regions, and made reference to specific countries within them. Hwang's fall from grace, which knocked him out of leadership contention, is described as something from which "researchers around the globe have had to back-track," but not as a blow to Asia writ large. The main competitors to the U.S. are identified as the United Kingdom (with its government spending $1.3 billion over ten years), U.S.-educated Chinese scientists returning to China, and Singapore's $400 million Biopolis (which can accommodate 2,000 scientists in government-funded and private and pharmaceutical labs). Sweden and India are mentioned in relation to hES cells for drug toxicity testing, and Israeli scientists are mentioned for their work on the intersection of genetics and hESC research. Japan is credited with an important breakthrough in reverse engineering mouse skin cells back into an embryonic state (human iPS cell lines had not yet been derived from human fibroblasts), and Finland is named with Singapore in providing the stiffest competition to the U.S. in terms of biotech development in the area.

The article, like earlier brain-drain articles, showed that people in many sectors were worried that the U.S. was falling behind, rather than providing short or long-term evidence that U.S. policy had in fact proved anti-competitive.[17] After listing the major competing countries, the article conceded that other nations, including Australia and Germany, had also struggled with ethical issues on the provenance and procurement of materials for hESC, and granted that there were reasons to be optimistic about U.S. progress in the field. The article ended by drawing attention to U.S. strengths but also to a rising threat from afar. One stem cell researcher was quoted as calling China the "sleeping giant" of stem cell research, but the article cited China's ability to test "experimental therapies directly on humans" in the context of freedom from the regulatory burden of clinical trials faced by U.S. researchers. While this treatment of China suggested that the nature of the pictured Asian competitive threat was beginning to change, China's testing of stem cell therapies on human patients did not lead to charges of under-regulated unethical and unproven science. That image of globally differentiated good and bad science would gain in salience, alongside a new stem cell research internationalism, within the next few years.

SINGAPORE, SOUTH KOREA, AND THE "EAST"

Stem cell research in the United States was always practiced in global science networks despite the nationalist and national competitive rhetoric that accompanied political and ethical funding battles of the first years of the new century. California, being a Pacific Rim state and benefiting from a robust inflow of Asian graduate students and postdoctoral fellows, looked to Asia for stem cell research collaboration and competition.[18] Nonetheless, California was not immune to the portrayal of an "East" with whom California stem cell researchers were in competition. Put crudely (and it often was), stem cell research was presented as patterning in one way in "the West" (that is, Europe and the United States) because of Christianity's objections to the embryo destruction involved in deriving human embryonic stem cell lines, while stem cell research in "the East" (that is, Asia) was presented as developing apace, in a relative regulatory oasis, because Asian religions did not recognize the pre-implantation embryo as a person and saw cloning as akin to reincarnation.[19]

As part of my research, I traveled in 2005 to two Asian countries with significant investments in stem cell research, Singapore and South Korea.[20] Both played an especially important part in the brain-drain imaginary during the time of President Bush's prohibition on federal funding, and the consequent heightened state, federal, and international competition, and during the rise and fall of Hwang Woo Suk. When I followed the above-mentioned image of stem cell science in "Asia" and "the East" to two of the countries in question, I found something rather different. Like the United States, both countries were developing nationally characteristic but different ways of carrying out stem cell research, profoundly intertwined with other countries, including the United States.[21] Rather than both fitting into some picture of the Eastern Other of Western stem cell science, they were involved at this period in developing national guidelines and were exploring whether or not there were regional identities to draw on in articulating national and international standards for stem cell research.[22]

What did South Korea's and Singapore's stem cell research efforts have in common and how did they differ from one another and from the United States? In what ways were they part of a larger "Asian" regional pattern to the emerging field of regenerative medicine, and to whom and for what purposes was that identity meaningful? Asia is the world's most populous continent, estimates of its share of the world's population running between 55 percent and more than 60 percent, depending on the source of statistics and the boundaries of Asia being assumed. This made it *prima facie* unlikely that there would be a single "Asian" identity. Also, many Asian countries had long since been leaders in different science and technology sectors, including biotechnology, and, increasingly, biomedicine. The exceptionalism characterizing the portrayal of Asian countries as leaders in stem cell research because the field was pushed to Asia by U.S. government policies, seemed less plausible considered against this background.

Interlocutors suggested to me that substantial similarities were to be found between Singapore and Korea on account of their both being "Asian Tigers." How might a shared "Asian Tiger" identity during restrictions on research in the United States have united Singapore and South Korea? The answer I got from various sources in the United States, Korea, and Singapore, and in various parts, can be reconstructed as follows: In view of the youth and breadth of the field, stem cell science combined the potential for truly significant scientific breakthroughs in diverse fields including molecular and cell biology, embryology, bioengineering, and biochemistry with tantalizing medical research possibilities and enormous promise clinically. The stakes for both symbolic and economic capital were thus significant. Encouraging economic growth and securing prestige are core functions of the modern nation state and the civic nation respectively, and are twin measures of "development." For nations investing heavily in innovation, stem cell science is a relatively low-hanging fruit, given that symbolic and economic capital are both likely outcomes of the research at this stage. Ethical barriers to research in the United States and several Western European countries—relating to the field's sourcing its basic materials from human bodies, often from embryos, and to its perceived potential to tamper with humanity—further served to make the field more genuinely competitive for other countries who may have fewer barriers to research. There is a unique opportunity to be a leader in the field, increasing national prestige and promoting government investment. For the Asian Tigers, regenerative medicine falls into the value-added category heavily promoted in the post 1997 Asian Financial Crisis era, relying less on exports and low wages and more on knowledge and research priority. Furthermore, stem cell research itself is turning out to be capable of being used to answer a diverse range of social questions about the body politic that vary from country to country. Like the closely related field of genomics, stem cell research can be put to very different ends and serve national ideologies. In sum, stem cell research is an ideal investment for Asian Tiger countries at this time.

I found, though, that comparing stem cell research in the two countries revealed convergences and divergences that belied the regional and economic parallels a shared label of Asian Tiger tended according to this reconstruction to suggest.[23] Singapore seemed to be engaging in stem cell research in a mode that I came to refer to as "knowledge society internationalism."[24] South Korea's pattern, on the other hand, seemed more continuous with the "developmental state" innovation for which it was famous,[25] and included a brief injection of a contested nationalistic sentimentality that made the Hwang Woo Suk story compelling worldwide. Stark contrasts evident in a key stem cell facility in each country reflected these differences. Despite the different paths of stem cell research in Singapore and Korea, however, I did find that "Asia" as an area identity was at least sometimes important in both countries, enabling both to articulate positions of geopolitical and scientific saliency in appropriate circumstances. I also found that, despite one strategy being rhetorically nationalist (the Korean one in this period) and the other being internationalist (the Singapore one), both countries' efforts furthered nationalistic visions that included a growing international profile for their countries, and an increased role for biomedicine in the post-Asian-crisis nation-state.

STEM CELL CULTURES

Singapore and South Korea are two of the four so-called Asian or East Asian Tigers (sometimes also called the "Little Dragons"), the others being Hong Kong and Taiwan, whose economies underwent rapid industrialization with high growth rates in the last four decades of the twentieth century. Singapore and South Korea share an experience with Japanese colonialism, but otherwise differ substantially in geopolitical terms. One is a tiny (with a population of approximately 4.5 million) primarily English-language city-state wedged between Malaysia and Indonesia (Singapore); the other is an ancient Korean-language-speaking nation of almost 50 million people, occupying the southern portion of a large peninsula, with a rich agricultural past. One was part of the British Empire; the other has housed U.S. forces since the Korean War. Singapore is more multi-ethnic; both have several large religious denominations, including Buddhism,

Taoism, and Christianity in Singapore, and Confucianism, Buddhism, and Christianity in Korea.

Both Singapore and South Korea are on the IMF's "Advanced Economy" list. In the 2007 UN Human Development Index, Singapore and South Korea ranked 26th and 27th respectively among nations in the world.[26] The standard account of industrialization in the Asian Tigers, as opposed to the standard account of Western industrialization, attributes its spectacular speed and magnitude to a combination of economic strategies and global relational starting conditions. The Tigers achieved this economic growth through the export of goods to rich industrialized nations; the leveraging of a period of low domestic wages relative to the countries targeted for export; a high level of state investment in national education systems including tertiary education; government mandated land reform to break aristocratic land tenure patterns; the use of tariffs and subsidies to control domestic spending; and investment in U.S. Treasury bonds to promote stability. The exact mechanisms accounting for the Asian economic miracle, and especially the reasons for the Asian financial crisis of 1997 are highly disputed, and are beyond my expertise. Based on my ethnographic data—what people say motivates them or explains things—it is probably reasonable, however, to designate 1997–98 a watershed in Korea, if not in Singapore. The IMF intervention, demanding neoliberal structural adjustments to control spending, raising interest rates, permitting financial institutions to fail, and putting an end to favoritism in securing foreign loans, contradicted some of the characteristics of the so-called developmental state. This was especially true in hard-hit South Korea, where the *chaebols* (large family-controlled firms with strong ties to government agencies) were used to garner oligopolistic political support and favoritism in terms of foreign loans. For Singapore to have weathered the crisis as well as it did, and for South Korea eventually to have exited the crisis, meant in both cases reduced reliance on high capital expenditure export driven growth, and a building up of foreign exchange reserves. The financial crisis marks a line in the sand after which the kind of innovation characteristic of knowledge economies, became more attractive. For Singapore, more than for Korea, investing in the new kind of bench-to-bedside biomedical research proved very much part of the answer as to how to do this. Korea, on the other hand, had a long history of putting science and technology at the center of its development by combining private-sector R&D with governance. This combination of private and public science and technology policy, and Korean *chaebols'* original focus on chemicals and electronics more than on the life sciences, meant that university research in the biomedical sciences stayed under the umbrella of the Ministry of Higher Education, Science, and Technology (itself a telling grouping of science and technology), and had a less clear path to industrial R&D. Similarly, Korea's deep commitment to education as development relies on sending elite students abroad, making education more of a qualification for taking a place in the private sector or in government upon return than in promoting an innovation space that encompasses Korean university labs and the private sector.

It is not uncommon to list cultural factors, especially "Chinese influence," among the traits that unite the Asian Tigers. In view of the different colonial and imperial relations and forms of nationalism pertaining in the four countries/territories over the last fifty years, and given the rapidly changing politics of China over this time and the relevant diasporas, and the consequent difficulties in attributing changing cultural Chinese influences to Chinese ethnicity or historical influence in the countries in question, I did not find it very helpful to my analysis. Ethnographically, I found a number of Singaporeans who attributed a role to Chinese ethnicity in producing growth and productivity. Among Koreans I heard relations with Japan and the United States mentioned, whereas "China" came up more as a synonym for remnants of the class system, often through its association with Confucian elites and regional dominance of some parts of Korea over others.

The economic policies listed above (export to the West, a period of low wages, state investment in education, land reform, and tariffs), produced and were in turn produced by a relatively high level of state authoritarianism or paternalism, combined with a high level of "economic freedom," and resulted in a trade surplus with highly industrialized countries and sustained high growth rates. In short, the pattern that in the West has become a stereotype of East Asia and its people—an extraordinary educational drive, high productivity, paternalistic government and social policies, while being friendly toward business —has a recent history and was produced in relation to the already highly industrialized world. The similarities between the historical time scales of the Tigers' industrialization and their resultant place in the global economic order might have led to a similarity between the countries' stem cell research efforts in this "post-Tiger" time.

I traveled to Korea in both 2005, at the peak of Hwang's fame and during the period of concern about an Asian brain drain, and again in 2008, after Hwanggate, after the achievement of induced pluripotency, and toward the end of George W. Bush's presidency in the United States. I also visited Singapore, the much-touted Asian brain-drain destination, in 2005.[27] My aim in 2005 was to compare the flagship laboratories of the two countries' respective stem cell research efforts, Hwang Woo Suk's laboratory in Korea and the newly constructed Biopolis in Singapore. The comparison, if it found variation within Asia, would suggest that a more sophisticated analysis was required: if two Asian Tigers were different, how could the entire of Asia (and the entire of the West) be accounted for by referring to Eastern versus Western religion? Once the Korean stem cell scandal broke, however, the comparison took on the additional aspect familiar from science journalism of attempting to account for the scandal: how/why did this happen in South Korea, rather than in the United States or Singapore? When I returned to Korea in 2008 it seemed that Hwang's rise and fall, though it probably could have happened anywhere, was narrated as being symptomatic of characteristics of South Korean stem differences between South Korean and Singaporean stem cell research, and gestured to the complexity and regionality of the life sciences in Asia writ large.

Hwang Woo Suk was a researcher and a member of the faculty at Seoul National University. SNU is considered to be the top university in South Korea, a country that valorizes education and is notorious for pushing its students to excel in exams. The summer of 2005 was the height of Hwang Woo Suk's fame in Korea and around the world for having reportedly succeeded in creating the world's first patient-specific human embryonic stem cell lines through the process of somatic cell nuclear transfer, or therapeutic/research cloning. During those months Dr. Hwang was one of a handful of the best-known scientists in the world: mediagenic, and possessed of an extraordinary knack for appearing and for narrating his work in a manner that was both modest and charismatic. Who could not be seduced by the story of hard work, Korean rural values, fame and honor from achievement rather than from being rich or a celebrity, and comprehensible and medically relevant scientific breakthroughs? In a time when the line between CEOs and scientists was becoming increasingly blurred in countries that were most aggressively adopting the innovation model in universities' life-sciences departments, this was the kind of scientific hero that the world longed for again. And in a country where educational achievement and recognition abroad confers symbolic capital, Koreans themselves both envied and fervently promoted Hwang, culminating in the government naming him Korea's "Supreme Scientist."[28] Even after Hwang's disgrace over the following months, many Koreans, including Korean women volunteering to be egg donors for his research, continued to believe in and support Hwang.[29] Many abroad, myself included, also wanted his research to be vindicated, for its medical and scientific promise, as a rejoinder to the threat of corruption posed by the over-corporatized innovation model, and as a welcome de-centering of a West-centered, English-language-dominated economy of scientific research.

Like others visiting Hwang's lab, I felt privileged to be there, witnessing what I believed was a great breakthrough.[30] Despite his ascendancy, Hwang was not without critics in Korea at the time. In fact, several people mentioned the casualties of what one person described to me as "the cult of Hwang," including the erasure of the contribution of his high-level collaborators (among them at least one prominent woman scientist who got little notice), and widespread envy of Hwang for his rise as well as his funding success in what is widely taken to be a zero-sum Korean science funding system. Pride may come before a fall, but so too does envy. Likewise, the rumors about ethical lapses in egg procurement were already widespread, and had reached me in California through my feminist networks before I left for Korea.[31] I was interviewed with a prominent Korean feminist bioethicist by a Korean newspaper during my visit, and my interlocutor openly voiced skepticism of Hwang's rise to the reporter. The charismatic nationalism, then, was all encompassing more in its emotional than its rational grip, and this in part accounts for its mythic exaggeration and the potential for fraud opened up by many parties wanting the phenomena to be real.

Visitors entering Hwang's lab were required to put on a protective lightweight jump suit over our clothes, to wear shoe covers, and to tuck hair into a scrub hat, as well as to pass through an air lock to decontaminate us. It is not entirely clear what functional value these protective practices held apart from creating a general environment of care: the lab was designed such that visitors entered down a central hall on one side of which was the human embryonic stem cell research and the other side of which was the training facility where the team worked on porcine ova.

The porcine side was the one through which visitors toured. Visitors were encouraged to imagine an isomorphism between the side they were visiting and the human embryonic stem cell side by the symmetry of the layout and by screens on the outside wall on the human side showing images of the work going on inside. It is not likely that anything being done on the porcine side would lead to human therapeutic biomaterials (cells that might be transplanted into a patient, for example), which might have made it important to protect tissue from contaminants carried by visitors. Nor is it likely that the Petri dishes of pig eggs, embryos, and stem cells posed a risk to us. On my visit, the rigorous contamination standards reminded those of us touring of sterile medical facilities on the one hand, and of silicon chip manufacturing on the other. The difficulty and mode of entering also seemed to mark the esteem in which Hwang and the lab were held. When I asked our guide about the egg procurement allegations, his answer was evasive and non-committal, making it clear that for me to ask further would be disrespectful.

Once inside the facility, we were taken around two adjoined rooms that together made up each of the stages of somatic cell nuclear transfer techniques and embryonic stem cell line derivation. The first station involved sorting and grossly preparing pig eggs, and the post-docs seated at that station literally had their hands in a large plastic basin of porcine ovarian material from the abattoir. Subsequent stations involved microscope and micromanipulation work representing the different stages of fertilization, incubation, and derivation of stem cell lines. Taken together, the lab resembled an artisan's workshop, with its guild members in training. Our guide, a lab member, described the time and dedication necessary to master each step, citing six months as the time it might take to become good at a particular micromanipulation skill. No one could move on to the next skill until the previous one was mastered. The highest standards of care were taken with the materials at each stage, conferring a profound embodied sense of the potential value of the materials and the techniques as they were transforming them.

This layout and these mores had deep resonances with the charismatic nationalist narrative of Hwang perpetuated by Hwang himself and others in the media. Hwang was already known for citing his rural roots as a livestock veterinarian and his rise from modest beginnings to supreme scientist. He credited the values of the Korean countryside with his work ethic, in a prelapsarian yet nationalist bucolic narrative appealing to media-consuming inhabitants of advanced capitalist cities around the world. His work ethic—a stated readiness to work 365 days a year, night and day if need be—was expressed in terms of there being not a second to lose to find life-saving and life-altering cures, in contrast to the capitalist "24/7" work ethic compelled by profit and self-interest. Above all, he had famously made the statement that he and his colleagues had succeeded in deriving embryonic stem cell lines from patient-specific cloned embryos because of national characteristics of the Koreans. Referring to the heavy metal chop-sticks Koreans typically use and the slippery foods they pick up with these chopsticks, he was frequently quoted around the world for claiming that this conferred a superior degree of manual dexterity in the Korean popula-tion as a whole. As such, the breakthrough in question, which required hitherto unattained manual skill, was naturalized to the Korean people, though some researchers in his lab were not Korean. Hwang's honor and glory was deflected from himself and onto the whole nation, only further fueling his rise. The ethos of care and guild-like apprenticeship of the lab reflected and exemplified these values of sacrificial hard work and pastoral humility. At the same time, the evident hierarchy and the presence of non-Koreans from other Asian countries such as Bangladesh (itself an interesting and important phenomenon, differing considerably from the kinds of science diasporas of Singapore's stem cell research) contradicted this picture.

The scientific payoff of Hwang's lab was not primarily progress in basic cellular and molecular research, advancing understanding of pluripotency and regeneration, or the advancement of tissue engineering prowess. Rather, it was being the first to succeed in applying a difficult veterinary technique to human cells for therapeutic benefit; namely, to get an enucle-ated human egg to begin dividing and differentiating after manually giving it the nuclear DNA from an individual afflicted with a condition in need of a cure. The promise of this technique was to be able to customize the DNA fingerprint of stem cell lines so as to treat individuals down the line with cells that bore their own DNA and so would not be rejected by the patient's immune system. While industrial scaling up seemed (and still seems) to be a daunting prospect, the therapeutic gains were apparent. The technique, known colloquially as "therapeutic cloning," is the one that was introduced to the world in Dolly the sheep almost a decade earlier. What Hwang's and colleagues' results appeared to demonstrate was that this procedure was possible in and viable for humans. It would have been a scientific first.

Hwang's credentials for carrying out the work were emblemized by his success in cloning the world's first dog: an Afghan hound named Snuppy, for "Seoul National University puppy."[32] Dogs, like humans and other primates, have notoriously difficult reproductive endocrinology, and so the feat was considerable. That Hwang's team should prevail in the race to clone human embryos and derive stem cells from them was plausible because of this existing expertise. But this difficult mammalian reproductive and embryological micromanipulation achievement far exceeded the veterinary cloning prowess from which it gained its credibility. The distance of this difference is evident in the activities to which Hwang had been consigned by my 2008 visit, namely the commercial business of pet cloning, while applying to re-enter mainstream science. With human cells, it appeared to be both proof of principle—patient-specific cells could in principle be made—and to have almost limitless therapeutic value—given that embryonic stem cells can give rise to almost all the cells of the body, each of us would have the possibility of an infinitely replenishable cellular repair kit in Hwang's hands.

The apparent achievement of Hwang and his colleagues was personal and national, the promise therapeutic and universal, and the ethos one of humility with glory, and meritocratic rather than economic success. This was a scientific priority race won by a charismatic scientist in university lab facilities, rather than the basis for a biotech start-up and the procurement of intellectual property. Once Hwang's team attempted to extend the achievement beyond Korea to found a worldwide stem cell hub, the rest of the world resisted, and joined Korean whistle blowers in uncovering the fraud that was revealed shortly thereafter.[33] One U.S. researcher who had got his name on one of the suspect publications despite perhaps having been less than a full co-author (illustrating the symbolic stakes and their international dimensions of scientific authorship) scrambled to disassociate himself and also became the subject of an investigation at his home institution.[34]

The ethos, the lab layout, the scientific goals, and even the iconic animal, were radically different in Singapore's Biopolis. The city-state of Singapore has been an independent republic since 1965, after brief periods of occupation, stewardship, or incorporation by Japan, Britain, and Malaysia respectively. Since then it has capitalized on the international potential of its English-language educational and legal system, and has dealt, through intense social planning, with extreme housing and land shortages. Biopolis is the name of Phase I of a huge custom-built biomedical research facility collectively known as One North, signifying Singapore's latitude and its aspirations to be a research and finance hub at the center of Southeast Asia.[35] Biopolis was built at the turn of the new century and displays a degree of social planning at once continuous with pre-existing waves of social planning and yet radically new. When I visited in 2005, Phase I of building had been completed and its buildings, named Chromos, Helios, Centros, Genome, Matrix, Nanos, and Proteos, were still being filled.[36]

Biopolis' newness for Singapore lay in part in the way in which it posits biomedical research as a way of life: somewhat as at a Silicon Valley company such as Google (but without the Peter Pan syndrome, as far as I could tell), one can get one's laundry done and socialize without leaving one's place of work. More significant for my argument here, though, is the turn-of-the-century mind meld between the public and private sectors that Biopolis' physical structure posits: two of the seven buildings are occupied by private-sector biomedical companies, the remaining five by the various biomedical research institutes of Singapore's Agency of Science, Technology, and Research (known as A★STAR). The seven buildings have a shared infrastructure and are connected by concrete "sky bridges."

Biopolis as a whole falls under the aegis of the Ministry of Trade and Industry, yet is made up of a mixture of public and private, educational and corporate entities: the Biomedical Research Council, the Science and Engineering Research Council, Exploit Technologies Pte Ltd, the A★STAR Graduate Academy, and the Corporate Planning and Administration Division. As A★STAR's very name suggests, it seeks excellence in scientific training, basic science research, and all stages of research and development together, encompassing what it calls a "full spectrum of R&D activities and graduate training."[37] This stands in contrast to the Korean (and U.S.) case.

This infrastructure for innovation has resulted in new areas of interdisciplinary biosciences research: of the five research institutes that make up the public part of Biopolis, the BioInformatics Institute, the Bioprocessing Technology Institute, the Genome Institute of Singapore, the Institute of Molecular and Cell Biology, and the Institute of Bioengineering & Nanotechnology, all were formed at the turn of the century except for IMCB, which dates back to the mid-1980s. A*STAR states as its vision "a prosperous and vibrant Singapore built upon a knowledge based economy," responsible for "fostering world-class scientific research and talent for a vibrant knowledge-based Singapore" and made up of "today's research scientists and future generation of aspiring scientists who dare to race with the world's best towards the very limits of modern science."[38]

I was taken around Biopolis by one of its researchers, just as at Hwang's lab. Being shown around Biopolis was rather like getting a real estate tour of an expensive new development, with the emphasis on filling the space with the right kinds of people.[39] Unlike at Hwang's lab, there were no sterility procedures; instead entry was regulated by security, and I was required to sign in and to wear a badge while on the premises.[40] The researcher showing me around emphasized the spaciousness, the layout, and the residents and research projects of the various lab spaces we visited. The most salient aspects of the tour concerned the residents, both animal and human, of the lab. High-status researchers from overseas headed up many of the populated labs. Europe, North America, and Asia were all represented, and not all of the lab heads were in residence year-round; some managed to keep academic positions elsewhere. Singapore universities are known for paying expatriate faculty members higher salaries than nationals, in a bid to lure faculty members with international credentials and reputations, so the pattern of having foreign born and/or trained heads of the labs had precedent.[41] Most of the students, however, seemed to be from Singapore.

If Snuppy the cloned Afghan hound was the totem animal of Hwang's lab and all it stood for, the zebrafish held this position at Biopolis. The zebrafish is one of the world's most commonly used research animals, because it is considered to be a good model of vertebrate development, including the basic biology of stem cell research, and reproduces quickly. A species with a long history as a valued tropical fish in Singapore, the zebrafish became central to Singapore's research infrastructure in the early 1990s in fish farming research. Biopolis' huge zebrafish research facility was established in 2004, and had apparently taken a certain amount of trial and error to set up.[42] It was still relatively new when I visited and was clearly a prize exhibit. The room was by far the largest animal-model facility I had ever seen. Far from Hwang's guild-like and heavily peopled lab, the zebrafish facility had no one in it when we visited. And far from being told how the hoped for scientific achievements of this space would be made possible by the national characteristics of researchers as occurred in Hwang's lab, the zebrafish facility was explicitly organized to allow researchers to choose which international style of zebrafish maintenance they preferred.

The fish tanks were organized into two sections, one kept in "the American style" and one in "the German style." I was told that there are two major schools of thought on establishing zebrafish populations for research: the American one, which enables food, light, temperature to be individually adjusted for each tank, as part of the experimental conditions, and the German one, which standardizes food, water, ambient temperature, and light. After some joking about the stereotypes involved—Americans and endless choice versus Germans and standardization—my guide told me that this layout enabled them to appeal to major overseas researchers, no matter their preferences or country of training/origin. In other words, the zebrafish were part of and partook in the qualities of the international real estate for science and technology design and organizational structure of the whole of Biopolis.

When I toured Hwang's lab, the ethics of human egg procurement dominated the unspoken space. At Biopolis, my questions about procurement of the eggs, embryos, or stem cell lines themselves needed for pluripotent stem cell research produced no anxiety in any of my interlocutors. Among the researchers I talked to, there was agreement that their biggest contribution was going to come from answering interdisciplinary foundational biological questions with biotech and pharmaceutical promise, and that most of the work in this direction could be answered through research on model species. Neither individual charisma nor contested ethics was in evidence, even though many of the major figures behind Biopolis were important and even notorious figures in Singapore society, and even though Singapore was actively in the process of writing and revising regulation to deal with the ethics of stem cell research, as were many countries around the world at the time.[43]

Biopolis' stem cell research, a central part of Biopolis' biomedical research activities, reflected Biopolis' organizational structure and was in line with the impression the researchers gave me of Biopolis' scientific priorities. Our guide, an A*STAR researcher, told me that a full spectrum of stem cell research, not just human embryonic stem cell research was encouraged, and that they were trying to benefit from having basic science research and translational and clinically relevant stem cell research all in the same site. He aligned Biopolis' fundamental approach to stem cell research as a "a spectrum" in the same manner as A*STAR described its overall mandate as covering the "full spectrum of R&D activities and graduate training." The recurrence of this metaphor during my visit was striking. Framing different kinds of research and different sectors by placing them on an imagined spectrum suggested that these activities and sectors naturally went together. This obscured just how original (and how unlike the situation in South Korea or elsewhere) it was to place these elements together in this way and that Biopolis' existence was original precisely for uniting these elements.

The use of spectrum metaphors also suggested that the spectrum itself was the real payoff, rather than the lab results of a single point on the spectrum. At Biopolis, the various spectra used to describe the place gave conceptual shape to the innovation hub being imagined. Not surprisingly, then, when I tried to find out what researchers and administrators were hoping would be the scientific payoffs of the stem cell activities at Biopolis (were there some equivalent goals to Hwang's team's attempts to make patient-specific stem cell lines, for example?), the answers were more about generating innovation, or at least a research environment indicative of innovation, in general. A success would not be judged by sensational news-worthy feats, but more by productive teams led by the right people and working in synergy—people whose output in publications and conferences would reflect and validate the value-added location. The practice of successfully luring major figures from abroad was jovially referred to as "serial kidnapping," and people reported on the percentage of the facilities in use to date. I was quoted facts and figures indicating productivity and hub like activity. More than being the first to do something the clinical relevance of which the public could understand, the three researchers to whom I talked at any length hoped to develop basic science research tools (such as a better understanding of the gene regulation of stem cell differentiation) that would fuel the onsite R&D chain and also would reinforce Singapore's reputation for knowledge infrastructure.

Singapore had set itself up as the central business and research and financial hub of Asia and beyond by seeking to recruit highly qualified international experts to train its young researchers, while Korea had moved much more slowly to open up its faculty positions to foreign researchers, and still sent a huge number of its elite students abroad to train. In the city-state of Singapore, business, education, research, and social planning were different facets of the same civic mission; at Biopolis, world-class stem cell researchers and other biomedical researchers were training a young largely Singaporean group of researchers to be the new citizens of an emerging knowledge society. In Hwang's lab, the emphasis was on Hwang's Koreanness, not on his lab's role in a knowledge society.[44]

In South Korea, much of university research, including much of the life sciences, falls under the Ministry of Higher Education, Science and Technology, and is an additional arena for displaying national educational competitiveness as a developmental strategy, rather than part of the powerful private but state protected industrial conglomerates that do most of the nation's R&D. The formation in 2008, under President Lee Myung-bak of the new Ministry of Knowledge Economy (MKE) represented an aspiration to more effectively integrate older developmental nation state practices with newer calls for financial deregulation, globalization, and innovation. The name of the new ministry was somewhat deceptive, however, as it replaced the older Ministry of Commerce, Industry, and Energy and remained separate from the Ministry of Higher Education, Science, and Technology.[45] Despite calls from some quarters to bring about a convergence between biological sciences and information sciences, and to move toward a knowledge society, university life-sciences research and translational biomedical research seemed harder to integrate into a knowledge economy in Korea than in Singapore because of their different pasts and goals. The protectionist and nationalist basis of Korea's previous decades of economic growth left an organization of research in Korea that was hard to change; only in 2008, with the establishment of the Ministry of Knowledge Economy, did Korea open up the directorships of its powerful research institutes to non-nationals.

When both Singapore and Korea reacted to similar global and regional trends by doing similar things, such as responding to the Asian financial crisis by moving somewhat away from high expenditure, foreign export manufacturing toward value-added knowledge economies, bench-to-bedside biomedical research such as stem cell research played a different role in the two places. Biomedical research, especially stem cell research, was a cornerstone of Singapore's effort in this regard, while Korea, later and less enthusiastic in this response in the first place, relied more on transforming its already highly advanced information and communication technology sector to accommodate World Bank demands, while stem cell research initially flourished in its national educational system. Arguably, then, Korea produced and then participated in the emotional nationalist drama of the fall of its "Supreme Scientist," because its university research labs were more part of its educational meritocracy, which was in turn central to Korea's development strategy, than they are part of its R&D sector.[46]

Singapore built and began to fill a facility devoted to a lifestyle of integrated research that embodied both the bench-to-bedside trajectory and the convergence of business, information, and biosciences, while taking care of all the living needs of its civic entrepreneurs. The prize of one was, or could have been, glory. After the fall of Hwang Woo Suk, the president of Seoul National University expressed it this way: "Most of us, in the name of national interests, exaggerated Dr. Hwang's research to make it an aspiration of the nation."[47] The prize of the other was its potential to be Asia's, if not the world's, "easiest place to do business," thanks to its stable if authoritarian legal, political, and economic environment.[48] Singapore led the way in regional intellectual-property law and finance reform, and signed a Free Trade Agreement with the United States in 2003.[49] Korea and the United States signed a substantial bilateral Free Trade agreement in 2007, but it was not ratified until 2011, and certain biomedical therapies patentable in the United States remained outside the scope of the agreement.[50] While Singapore continued to pay foreign faculty members more than its nationals and to recruit superstars from prestigious universities overseas, Korea saw one of its own nationals become a household name around the world. There were, as ever, profound regional and local differences in how "the same science" is enabled, practiced, and understood. In such small things as lab layout and mores, and in scientific strategies, strengths, and weakness, the two flagship labs represented very different ways of being Asian Tigers of stem cell research and the brain-drain imaginary. There was no unified "Asia" or "East" in evidence. Researchers in both countries claimed to be operating within Asian geographic, historical, and economic patterns.

As 2007 turned to 2008, the upcoming U.S. presidential election looked likely to bring about a change of administration that would be more supportive of federal funding for embryonic stem cell research. The joint publication by Japanese and American teams of successful human induced pluripotent stem cells lines, presaged the imminent normalization of human pluripotent stem cell research. A reverse brain drain resurfaced in the news. In May 2008, Alan Colman (member of the team that cloned Dolly the sheep) took up a position at King's College in London. Journalists writing in the United States, though acknowledging that Colman would still be spending a proportion of his time in Singapore, described his return from Singapore as, for example, "only the latest high-profile researcher to announce he is leaving," saying that it "does not bode well for Singapore's aspirations to dominate the stem-cell research field."[51] In Singapore, A*STAR described the move by saying that "Dr. Colman will build bridges between Singapore and U.K. stem cell research communities," but it was described in a Singapore blog site as "dealing another blow to the city-state's biotech ambitions."[52] Lee Wei Ling, a member of Singapore's first family and the director of its National Neuroscience Institute, criticized the policy of luring high-flying overseas researchers for part-time jobs at huge salaries and she spoke out in favor of indigenizing Singapore's science education and future science workforce.[53] Colman's stay at King's College was short, however, and he was back in Singapore full time by the fall of 2009.[54] By 2008 the era of international stem cell research collaborations had begun in earnest.

STEM CELL INTERNATIONALISM VERSUS STEM CELL TOURISM

By 2008, a new stem cell internationalism was evident. The nationalism and the competition that were so important to the story of the politics and ethics of pluripotent stem cell research in the first decade of its existence were being supplemented by a growing internationalism. This expressed itself through the rise of international professional stem cell research organizations and related professional biomedical organizations, regulatory attention by international organizations, and a growing countervailing rhetoric within California and elsewhere of the value of international collaboration. In the background, every aspect of the science, from education to professional meetings, from patenting laws to venture capital, and from material-transfer agreements to research and publishing collaborations, continued to reflect the traffic between different parts of the world that is the hallmark of modern science.[55]

TABLE 4.2
Human pluripotent stem cell research takes root: comparison of United States, South Korea, Singapore, and the United Kingdom.

	United States	United Kingdom	South Korea	Singapore
Example of pioneering hESC scientist, science system (the face of stem cell research)	James Thomson (first HESC lines, 1998; co-pioneer, iPS cell lines, 2007): genetics and mammalian development	Sir Martin Evans (first mouse ESC lines, 1981): genetics and mammalian development; also known for knockout mice	Hwang Woo Suk (first cloned dog; claimed falsely first SCNT lines) veterinary science; patient-specific lines	Foreigners and nationals, e.g. Alan Colman and Arif Bongso; cancer research; stem cell culture medium; stem cell signaling
Other important factors	Biotech, human genome project infrastructure; leftover IVF embryos; venture capital	IVF, PGD, stem cell banking pioneers, infrastructure for handling eggs and embryos	State investment in technology for development and adulation of successful scientists	Interdisciplinary focus on regenerative medicine tools and as business
Characteristic organism in media	Snowflake children	Dolly the sheep	Snuppy the dog	Zebrafish
Characteristic institution during start up of human pluripotent stem cell research in each country	Geron, biopharmaceutical company which owns or licenses telomerase and hESC patents, and cancelled hESC trial	HFEA, the Human Fertilization and Embryology Authority, which regulates human embryos and gametes, and U.K. Stem Cell Bank	Hwang's erstwhile lab at Seoul National University, SNU, for which Snuppy (SNU + puppy), was named	Biopolis, a biomedical research planned community that mixes public and corporate research laboratories
Summary	Innovation; priority; abortion debate; pro-cures; biopharmaceutical markets; patents	Embryology and its genetic control; fertility treatments; bio-banking; research cloning; regulation	Science nationalism; development through support for technology	Basic flexible regenerative sciences; interdisciplinarity; international hub

California Code of Regulations Section 100080, titled Acceptable Research Materials, was amended several times in the first five years of CIRM's activities. It had quickly become clear that the numbers of high-quality pluripotent stem cell lines available for research could be increased by allowing CIRM to fund research with stem cell lines derived under other countries' approval regimes, if those countries were deemed to have equivalently rigorous ethical and scientific standards. As they were approved, Section 100080 a) 1) listed lines approved by the U.S. National Institutes of Health, and those derived in accordance with the passage of Proposition 71. It also listed lines from the following countries and institutions:

United Kingdom: lines deposited in the United Kingdom Stem Cell Bank, and lines approved for use by a licensee of the Human Fertilisation and Embryology Authority
Canada: lines derived in accordance with the Canadian Institutes of Health Research Guidelines for Human Pluripotent Stem Cell Research under and application approved by their National Stem Cell Oversight Committee
Japan: lines derived in accordance with the Japanese Guidelines for Derivation and Utilization of Human Embryonic Stem Cells.

At a CIRM-sponsored meeting I attended in 2006, a debate took place as to whether or not a set of standards could be developed according to which any countries' lines that complied could become eligible for funding, rather than having to list countries in the regulations individually. One scientist pointed to the waste of time and money, and the potential consequent bottleneck for research, involved in the process CIRM was required to go through to amend the regulations every time a new country's own standards for acceptable derivation were recognized by CIRM. While several people seemed to like the idea of a universal standard in principle, most thought it was prudent to be specific, so as to have explicit control over the lines around the world that could be used in stem cell research funded by CIRM. At this stage, a geography of the acceptable circulation of stem cells, based on certified ethical and scientific standards, was being accreted, country by country, in CIRM's policies.

Australian biologist and IVF and embryology pioneer Alan Trounson became CIRM's president in January 2008. The caution around international collaboration evident in the process by which countries' stem cell lines were listed in Section 100080 a) 1), and the earlier nationalist competitive "brain drain" rhetoric notwithstanding, Trounson made the forging of international collaborations one of the things for which he became known. Almost immediately he started signing collaborations between CIRM and other countries. In one sense, this was continuous with the competitive stance taken by California around Proposition 71. CIRM, in signing research collaborations with other countries' government science funding agencies, was again acting like a country, or at least the cutting edge of the country, rather than like a single state in a much larger country engaged in stem cell research. It leapfrogged the federal government, with, as far as I could discover, no resistance from the federal government, and no objections from the entities with which it was forging new bonds. But in that these collaborations relied on a vision where international connectivity would increase, rather than decrease competitiveness, this was the beginning of a new geography of stem cell research.

CIRM's webpage describing its "Stem Cell Research Collaborative Funding Agreements," addresses the question "Why form collaborative partnerships?"[56] as follows:

The California Institute for Regenerative Medicine funds research carried out in California. However, excellent stem cell science is taking place worldwide. Our collaborative funding relationships facilitate work between Californian scientists and their innovative colleagues around the globe. . . . Collaborative funding . . . creates a critical mass of excellence across a wide range of specialties worldwide. For CIRM grantees, the program expands the number of scientists with whom they can work and mitigates the geographic limitations imposed by CIRM's funding.

The agreements would overcome what had come to seem a geography of limitation, resulting from the public state-based nature of CIRM grants. The original three countries included in the state constitution under Section 100080 a) 1), Canada, the United Kingdom, and Japan, were all early co-signatories of Memoranda of Understanding (MOUs), along with Trounsen's home state of Victoria, Australia, and Spain. By the end of 2008, CIRM had collaborations on the books with Canada's Cancer Stem Cell Consortium (June 2008); Australia's state of Victoria (June 2008); the United Kingdom's Medical Research Council (September 2008); the Japan

Science and Technology Agency (November 2008); and Spain's Ministerio de Ciencia e Innovacion, and the Andalusian Initiative for Advanced Therapies (December 2008).

One might have predicted that the need for CIRM collaborations would be reduced once Barack Obama took office as president, because of the improved ease of doing research with federal resources. Certainly, CIRM's leadership appreciated the new NIH Guidelines and, in particular, praised their internationalization. Trounson said of the new NIH guidelines: "The NIH showed foresight by specifically laying out a procedure for how lines derived outside the United States may be included in the registry. The revised policy recognizes the global aspect of stem cell science and supports existing international collaborations."[57] Nonetheless, CIRM continued to forge new collaborations throughout Obama's first term. It signed MOUs with the state of Maryland in 2009 and the New York Stem Cell Foundation in 2010, before establishing a formal collaboration with the federal government through the National Institutes of Health in September 2011. CIRM signed MOUs to permit grant applications to CIRM with overseas collaborators who could share the work and the costs of research, with more countries and agencies in Europe and Asia-Pacific, and by 2012 had added countries from Latin America to its North American collaborators. MOUs were signed with Germany's Bundesministerium für Bildung und Forschung (September 2009); China's Ministry of Science and Technology (October 2010); India's Institute for Stem Cell Biology and Regenerative Medicine, under the Ministry of Science and Technology (December 2010); France's Agence Nationale de la Recherche (January 2011); Australia's National Health and Medical Research Council (April 2011); Scotland's Scottish Enterprise (August 2011); Argentina's Ministry of Science, Technology and Productive Innovation of Argentina (March 2012), and Brazil's National Council for Scientific and Technological Development (March 2012). Singapore and South Korea were noticeably missing from the list, although Trounson himself had strong links to Singapore, having been awarded a medal there in 1995 and having been part of the launching of ES Cell International, housed in Biopolis. The absence of any agreements with South Korea probably referenced the Hwang scandal, and perhaps reflected Korea's innovation system and residual nationalisms, but Korean stem cell scientists continued to be active on U.S. campuses and at international conferences.

CIRM signed one of its first MOUs with a major U.S. disease advocacy group, the Juvenile Diabetes Research Foundation, in August 2008, underlining the core role of patient advocate organizations in CIRM's identity.[58] In February 2012, it signed a MOU with the private Keystone Symposia group, a Colorado-based 501(c)(3) nonprofit interdisciplinary molecular and cellular biology conference conceptualizing and hosting organization with roots in California, enabling the Keystone Group to contribute monies to host a conference partly funded by a CIRM grant held by Stanford University researchers. From Juvenile Diabetes Research Fund funding to get around federal restrictions on procurement of embryonic cells in 2008 to signing a collaboration with Keystone in 2012 so as to leverage more than one source of funding to host a professional meeting, CIRM's stem cell internationalism had moved from cost sharing the funding of controversial research to helping stage international meetings of minds.

FLAT WORLD?

The decade after President Bush's ruling on stem cell research saw the rise of several forums and alliances that formed to promote the internationalization of stem cell research. The International Society for Stem Cell Research (ISSCR) was formed in 2002 to bring together stem cell scientists from around the world to share the latest scientific information and to work toward international standardization and harmonization for the regulatory, ethical, intellectual-property, and material-transfer aspects of the research.[59] Beginning in 2003, the ISSCR issued regular position statements, usually supporting political initiatives to regulate and fund stem cell research and protesting efforts by governments or the United Nations to restrict stem cell research or its funding. It also maintained a "For the Public" site that provides high-quality educational information on stem cell research and its clinical translation. The basic premise of the ISSCR was that science is by its very nature international, and that the best research occurs in an environment of responsible conduct and open exchange of ideas.[60] This carried its own politics, as evident from the educational materials the organization produces, its sponsorship of student stem cell societies and its intervention through its position statements. Nonetheless, it was importantly apolitical in another sense, eschewing the narrow nationalisms that foster secrecy in science.

This "flat world" of collaboration and internationalism was not the whole story, however. As the Korean and Singaporean Asian stem cell Other and its associated tropes of international competition for scientific prestige and brain drains began to recede in the U.S. stem cell imaginary, a second Asian Other was rising. This one centered more on India and China, along with certain Eastern European and Central and Southern American countries, than on the Asian Tigers.[61] By the end of 2008, the International Society for Stem Cell Research had published *Guidelines for the Clinical Translation of Stem Cells.*[62] This document, prepared by the ISSCR Task Force on Clinical Translation of Stem Cells, was a template for everything that had been learned to date in international stem cell regulation efforts (I will return to its rehearsal of animal-model science verification and regulation in chapter 5).[63] But it also raised a flag that regulatory gradients were being exploited to market unproven treatments directly to desperate patients. The overblown therapeutic promise, the "hype and hope" integral to the pro-curial package, had moved from being a way to attract voter assent for funding to being the basis for a feared exploitation of patients. This concern with rogue science and fraudulent claims about cures got expressed in general terms, but was also described as being particularly worrisome where patients had to cross borders in the pursuit of these unproven therapies. The flat world, it turned out, had topology.[64] The report said:

> The ISSCR recognizes an urgent need to address the problem of unproven stem cell interventions being marketed directly to patients. Numerous clinics around the world are *exploiting patients' hopes* by purporting to offer new and effective stem cell therapies for seriously ill patients, typically for large sums of money and without credible scientific rationale, transparency, oversight, or patient protections. . . . The marketing of unproven stem cell interventions is especially worrisome in cases where patients with severe diseases or injuries *travel across borders* to seek treatments purported to be stem cell-based "therapies" or "cures" that fall outside the realm of standard medical practice.[65] (emphases added)

This passage emphasized the risks to hopeful patients when they are subject to unscrupulous marketing and unsafe or inefficacious stem cell interventions. It did so by referencing, without explicitly citing, two prominent debates about biomedicine in the United States in 2008. The first debate, evident in the phrase "unproven stem cell interventions being *marketed directly to patients,*" concerned the hierarchy of authority in medicine, and whether or not and to what extent it was acceptable for bio-medical products not covered by regulations prohibiting it, to be marketed directly to consumers, without going through government regulators and physicians. In 2008, this debate had been raging in another quarter, direct-to-consumer (DTC) genetic testing. Amid a patchwork of Food and Drug Administration and state regulations, and against a background in which the Federal Trade Commission had previously advised "a healthy dose of skepticism" for "at-home tests" on behalf of consumers, the American Society of Human Genetics (ASHG) issued guidelines on health-care-related direct-to-consumer marketing in late 2007.[66] In the ASHG's report, the tests were not referred to by their at-home-ness, but by their marketing directly to the consumer; it was false marketing (exploiting the patient), rather than paternalism (the state and the doctor know better) against patient empowerment, that was to become the problem with the tests upon which their ethical and epistemic failures would be blamed. The ASHG recommendations fell into three categories, each of which rein-stated a pillar of authority and hierarchy and accreditation, now that the problem had been localized.

The first principle, "Transparency," called for adequate information and regulation for patients to be able to exercise robust informed consent and have their privacy protected. The second principle, "Provider Education," sought to reinstate health-care providers between product and patient; providers and their professional organizations were to educate themselves so they could protect their patients by interpreting the analytic and clinical validity and efficacy of the tests. The final principle, "Test and Laboratory Quality," sought to engage all the relevant government regulatory bodies to ensure the tests were safe and efficacious. Flattening the health-care landscape through direct-to-consumer marketing was portrayed as unsafe, ineffective, and exploitative (with good reason—several tests and the meth-odologies upon which they rely had been shown to be at best mislead-ing[67]), rather than empowering of patients. I could not have been alone in reading the ISSCR guidelines against this defense of the need for regula-tion and the medical profession as quality-control agents between con-sumer and biomedical product. Stem cell treatments were emerging as just as much of a regulatory mishmash as DTC genetic tests, and with poten-tially much more serious consequences because of the intrusive nature of the "therapies" being peddled.[68]

The second debate referenced in the ISSCR guidelines was that surrounding the growing phenomenon of medical tourism, including organ
transplants, assisted reproductive technologies, cosmetic and plastic surgery,
and dentistry.[69] In 2008, the American Medical Association issued a list of
nine principles directed toward those who "facilitate or incentivize medical
care outside the U.S."; although it addressed only those who traveled as
patients, from a U.S. point of view, and not those traveling or staying in
place to provide service, or those traveling within or to the United States,
it was a powerful patient bill of rights, seeking to ensure safety and quality
for medial travelers.[70] This need is brought into urgent relief when read
against the relevant part of the ISSCR guidelines, concerning "patients
with severe diseases or injuries," who "*travel across borders* to seek treatments
purported to be stem-cell-based 'therapies' or 'cures' that fall outside the
realm of standard medical practice."[71]

Like other medical tourists, patients seeking stem cell treatments will
sometimes be prepared to go wherever the procedures are offered, if they
feel they have run out of options.[72] The American Medical Association's
apparently neoliberal, patient-empowering guidelines notwithstanding, the
world of medical tourism is no more a flattening of the world than the
introduction of direct-to-consumer marketing.[73] Governments receiving
medical tourists often promote or turn a blind eye to medical tourism
because it earns foreign exchange, enhances the national economy in direct
ways and through associated tourism expenditure, and because it advances
the knowledge economy with its attendant educational, developmental, and
civil-society benefits. In other words, in the eyes of governments, medical
tourism can be tourism rather than medicine. Receiving government
support for medical tourism does not necessarily benefit the traveling
patients or the populace, however, and may harm patients and lead to a
deterioration of locally available health care.[74] Sending governments also
promote or turn a blind eye to medical tourism. It can bring down health-
care costs to the government or to the private sector, relieve pressure on
over-extended hospitals and clinics, and offshore some of the labor and
clinical trial costs for pharmaceutical and biotech companies developing
therapies. And third, service providers of medical tourism follow their own
gradients. Practitioners (doctors, students, hospital administrators, non-elite
medical personnel, and the biolabor reserve force) of medical tourism
engage in migrations that mark out knowledge and credentialing gradients

around the world as well as labor gradients and immigration aspirations. In turn, medical labor gradients include push factors such as escaping poverty or armed conflict or ethno-racial or gender or sexuality violence, engaging in work as a medical research subject rather than something more risky (such as prostitution), or to feed a family, or starting an immigration path by getting a foot in the door and a professional qualification in another country. Stem cell treatments and regenerative medicine will surely become a focus of medical tourism in the years to come.

In sum, stem cell research was invoked as displaying several distinct geopolitical patterns in this period. On the one hand, there was a kind of international republic of stem cell scientific knowledge, and international standardization and harmonization of ethics, backed by borderless patient advocacy and its civic epistemology. This all suggested an increasingly horizontal, international world. At the same time, transnational and domestic labor (including scientific labor) and tissue markets, as well as radical differences in access, affordability, and health outcomes suggested a recalcitrant vertically stratified transnational stem cell world. New kinds of academy-industry alliances around human cellular therapies and the flow of capital signaled globalization. Yet other elements suggested the growing importance and perceived threat of emerging economies in countries such as India, China, and Brazil, to stem cell research, and of the quasi-post-socialist landscape in Eastern Europe and China to stem cell medical tourism and science diasporas. Stem cell medical tourism typically entrained little that looked like the neoliberal vacation of common understandings of medical tourism as a beach holiday with health or cosmetic interventions, but as stem cells are used more and more in cosmetic procedures, that will no doubt become a common stem cell geography, too.[75]

By 2012, the construction from the United States of an imagined Asian regulatory paradise, its crisis around the Hwang scandal, and its conversion into a problem about good versus bad science with a different, often Asian-centered imaginary, had come full circle; once again, the United States and its others could not be as easily separated as these geographies seemed to imply. In January 2006, the editor of the influential *American Journal of Bioethics*, Glenn McGee, had requested that Drs. Insoo Hyun and Kyu Won Jung, authors of a paper the journal had published on the informed-consent procedures for egg donors for Hwang's SCNT research, withdraw their paper once it became clear that the procedures had not been followed.[76] Hyun, co-chair of the ISSCR *Guidelines* task force, and Jung, a scholar of Korean bioethics, were not themselves accused of any wrong-doing, and each of them played a part in the next phase of international stem cell ethics.[77] Six years later, McGee felt it necessary to resign from the editorship of the *American Journal of Bioethics* amid allegations of possible conflicts of interest during his transition to the private sector and a position with the U.S.-based stem cell venture CellTex. CellTex was collaborating with Korea's RNL Bio in performing adult stem cell banking and treatments that had not been permissible in Korea since the Hwang scandal, and CellTex had been criticized for the absence of safety and efficacy data.[78] International stem cell circuits continue to evolve, including stunning advances in science and emerging ethical quagmires, that cross borders and co-implicate researchers from different countries. A rhetoric of East versus West, or good versus bad science is much too crude and inaccurate (and, to many, offensive) to capture the geopolitics of the science or the ethics. It is hard to know what will come to seem, as if all of a sudden, to have been bad science or unethical practice. This probably means, except in certain cases of outright fraud or unmitigated greed, that it is crucial not to over-emphasize winners and losers, or to think that anyone can definitively separate ethical and unethical practitioners. However, that caution against emphasizing winners and losers, the good and the bad, does not mean that the process of incorporating perspectives and developments in science and ethics into regulations and norms should be halted. To the contrary, these moral panics and epistemological reorientations, and the resultant ecdyses of practices that have compromised ethics and science, are essential to the very possibility of good science. Pushing through these events, and sorting out what will make practice better in the future and what parts of the research work with what compromises, is what will have to be done, because there is no one geography of good and bad science.

III   THINKING OF OTHER LIVES

5 A FORWARD-LOOKING STATE: ON PUBLIC DONATIONS AND RECIPROCITY IN CALIFORNIA'S STEM CELL PROPOSITION

TWO PUBLIC GIFTS AND THEIR TAKING

The preamble to the 2004 California Stem Cell Research and Cures Act had all the hallmarks of a forward-looking statement. In the world of business, a forward-looking statement is a promissory account of a company's mission that stokes investors' enthusiasm for a company in a way that is reliable enough to contribute to a decision to invest, while being explicitly exempt from being an accurate account of what turns out later to have been true.[1] According to the Farlex Inc. Financial Dictionary, "A forward-looking statement makes certain assumptions. . . . As a result, forward-looking statements are subject to revision, as reality may not match the assumptions. A forward-looking statement is nonetheless useful for making certain decisions about a company's future, such as whether or not to expand operations."[2]

Companies in the United States are assured a "safe harbor" for forward-looking statements that turn out not to have been accurate, as long as word choice makes it clear the statements were forward-looking, they were made in good faith, and they were helpful in understanding the investment risks involved. Proposition 71 did all these things. It stoked enthusiasm, as witnessed by the easy majority with which the proposition passed. It was, I have every reason to believe, written in good faith, while being reliable enough to provide guidance as to the potential for market success for those investing in the research (the public, in this case, but also many private donors). And the claims were hedged with characteristic forward-looking adverbial expressions and conditionals, and forward-directed prepositions and verbs. In its "Findings and Declarations," the text stated the following (emphases are added):

Recently medical science has discovered a new way to attack chronic diseases and injuries. The cure and treatment of these diseases *can potentially be accomplished* through the use of new regenerative medical therapies including a special type of human cells, called stem cells.

These life-saving medical breakthroughs *can only happen if* adequate funding is made available to advance stem cell research, develop therapies, and conduct clinical trials.

About half of California's families have a child or adult who has suffered or will suffer from a serious, often critical or terminal, medical condition that *could potentially be treated or cured* with stem cell therapies. In these cases of chronic illness or when patients face a medical crisis, the health-care system *may* simply not be able to meet the needs of patients or control spiraling costs, unless therapy focus switches away from maintenance and *toward* prevention and cures.

Unfortunately, the federal government is not providing adequate funding necessary for the urgent research and facilities needed *to develop* stem cell therapies to treat and cure diseases and serious injuries.

This critical funding gap currently prevents the rapid advancement of research that *could benefit* millions of Californians.[3]

The California constitutional amendment that was enacted by Proposition 71 had fewer "forward-looking" clues, but those familiar with the genre would still have noticed language that somewhat tempered the promise of cures in a fixed time frame. For example:

The institute shall have the following purposes:

(a) To make grants and loans for stem cell research, for research facilities, and for other vital research opportunities to realize therapies, protocols, and/or medical procedures that will result in, *as speedily as possible, the cure for, and/or substantial mitigation of,* major diseases, injuries, and orphan diseases. . . .

On the one hand, the "forward-looking statement" aspect of Proposition 71 made a great deal of sense, insofar as the point was to regenerate California's economy while seeking to launch regenerative medicine in the state. Consider the ways in which the proposition proposed to protect and grow the state's economy:

Section 3, Purpose and Intent: Protect and benefit the California budget: by postponing general fund payments on the bonds for the first five years; by funding scientific and medical research that will significantly reduce state health care costs in the future; and by providing an opportunity for the state to benefit from royalties, patents, and licensing fees that result from the research.

Benefit the California economy by creating projects, jobs, and therapies that will generate millions of dollars in new tax revenues in our state.
Advance the biotech industry in California to world leadership, as an economic engine for California's future.

The initiative was geared to getting a competitive edge in the nascent regenerative medicine field, creating jobs, generating tax revenue, launching intellectual-property-based income streams, and reducing the costs to the state of health care. In short, it was a business, as well as an ethical, proposition for California voters and taxpayers. On the other hand, however, voters did not necessarily recognize the genre of the forward-looking statement, and a publicly funded initiative, especially one to do with health care and disability, was not necessarily the place to behave so much like a business. At the least, two aspects of the comparison of Proposition 71 and a forward-looking statement call out for further discussion. While businesses are bound by laws that stipulate that they must file information that is—unlike forward-looking statements—subject to reality, and that must assess a wide range of investment risks, the drafters of Proposition 71 had no such obligation. Even voters accustomed to balancing forward-looking statements with, for example, forms filed in compliance with the rules of the Securities and Exchange Commission in weighing their investment options (surely a tiny minority of voters) would not easily have unearthed the anticipated risks. At best they could have looked to the campaign's naysayers, each of whom had his or her own reasons for opposing the proposition, but the latter's statements were not necessarily in line with what the drafters of the proposition themselves would have seen as the most relevant risks of the proposition as it was crafted. Another imperfect parallel between Proposition 71 and a forward-looking statement lay in the ways in which the public was investing in the research, and the ways in which they could thus expect to be rewarded. The voters were not simply financial investors in stem cell research who could expect investor value to be upheld. As I explore below, the public was also enlisted as a donor of bodily tissue to the research. The ways in which return on these different kinds of investment was to be measured for funding and for tissue donation were underspecified and contested.

Proposition 71, then, while not exactly a forward-looking statement about the state's proposed investment in the stem cell industry, did portray California as a forward-looking state pushing its innovation economy forward through this pro-curial[4] field of biomedical research. To some extent it was possible (and still is) to find out the kinds of risks that the state taxpayer needed to weigh against the field's promise when voting to fund the research by looking at the SEC filings for nascent companies in the state, as has been done to understand the risks inherent in genomics speculation.[5] However, because of the second failure of the parallel between the Proposition and a forward-looking statement—that taxpayers are not just investing funding but also investing bodily tissue, and are expecting not only several different kinds of economic benefits but also health-related returns from both kinds of investment—pluripotent stem cell companies in the state shed light on what might be expected of a forward-looking state as much for the ways in which they are out of sync with what Proposition 71 promised. Consider the case of one corporation, which is instructive in this regard.

This company had not, at the time of my interview or of writing, received grants from CIRM for its main technology based on deriving pluripotent parthenogenetic stem cell lines from donated oocytes, although it was in theory eligible for CIRM funding and had applied in the past and was intending to do so in the future. At least one employee had previously received CIRM funding for training.[6] The technology was ineligible for federal funds because parthenogenesis was still considered in federal law to be a form of embryo cloning.[7] This technology, however, by avoiding the need to destroy embryos (see table 5.1) yet keeping many of the physical advantages of embryonic as opposed to induced pluripotent stem cells, held a great deal of promise.[8]

Aside from its success in deriving pluripotent stem cell lines from parthenogenesis, a wholly owned subsidiary of this company announced in June of 2012 that it had developed a way to modify human stem cells using engineered proteins instead of viruses or chemicals, making them more likely to be therapeutically safe. It also held licenses to important stem cell intellectual property. These developments suggested that the company had promising pluripotent stem-cell-based technology in development. It was, in other words, the kind of company that was trying to develop promising pluripotent stem cell technologies in California that could not receive federal grants, and that was ready to take on the early stages of R&D for the field in the state. Such promise, however, did not translate immediately to financial viability for the company. And it depended on getting right the terms of procurement of the women's eggs that were needed for the technique.[9]

Unlike the text of Proposition 71, the company's finances and business model, and its Securities and Exchange Commission filings, made the risks inherent in investing in pluripotent stem cell research all too clear.[10] Its Securities and Exchange Commission Prospectus filed pursuant to Rule

TABLE 5.1

Types of human source material for pluripotent stem cell research, and frame of donation.

Type of pluripotent stem cells	Human source tissue / cell type(s)	Frame of donation
Embryonic stem cells	Eggs and embryos are procured in a reproductive context and are usually "supernumerary" to reproductive needs, though eggs can be from egg sharing scheme; research donation of eggs rare. Sperm donation can be regular, onerous paid piece work, as well as part of reproductive scenario.	Embryos, adopted or given to research, frozen or discarded, are precious and hard to turn into waste. Sperm is sold screened and frozen, and differentiated in reproductive context. Eggs are risky to extract and scarce in reproductive and research contexts, but the labor, and in reproductive contexts the differentiated value, can be highly compensated; usually extracted and donated "fresh," but immature and frozen egg donation growing.
Induced pluripotent stem cells[a]	Somatic cells, often fibroblasts or fat cells, acquired by commercial companies from consented medical waste appropriation, or fresh skin punches or liposuction	Fresh tissue collection and tissue that would otherwise be medical waste and discarded, increasingly needs patient and clinic consent but can then be processed and sold by commercial entities or used by scientists
Parthenogenetic or parthenote stem cells[b]	Egg donation as side product of self or other's reproduction; with or without somatic cell DNA addition	Supernumerary, unfertilized, or shared eggs gifted with compensation for reproduction, but usually not for research

TABLE 5.1 *(continued)*

Type of pluripotent stem cells	Human source tissue / cell type(s)	Frame of donation
Somatic cell nuclear transfer stem cells[c]	Egg, enucleated, and somatic cell nucleus donation	See above.
Placental and cord blood stem cells[d]	Products of pregnancy tissue from hospitals, tissue banks, and abortion clinics; commercial company processes material; proportion of profits may fund clinics	Consent from "mother" even in abortion cases,[e] and to a lesser extent from "father"; tissue would otherwise be medical waste and discarded

a. Human iPS cells are not currently identical to hES cells; the former seem to have an epigenetic "memory" evident in methylation differences. Because of this, and because of unknown effects of the necessary reprogramming of gene expression patterns to achieve pluripotency, the clinical future of iPS cells is still in doubt. See, e.g., Lister et al. 2011.

b. Parthenogenetic or parthenote human pluripotent stem cells are derived from unfertilized women's eggs that are induced to divide as if fertilized to blastocyst stage, from which the inner cell mass can be extracted. See, e.g., Revazova et al. 2007. Recently, it has been shown to be possible to combine this method with the addition of the genetic material from a somatic cell, suggesting a way access the histocompatible promise of somatic cell nuclear transfer, despite the technical hurdles in performing the latter when the egg is enucleated. See Noggle et al. 2011.

c. This procedure has yet to be successful with human cells. See Wilmut et al. 2002 for an early summary of the challenges and procedures.

d. For pluripotent umbilical cord blood stem cells, see Harris and Rogers 2007. For amnion-derived pluripotent stem cells, see Miki and Strom 2006. Certain fetal stem cells may also have pluripotent properties.

e. Consent to procure the products of pregnancy, including fetal tissue, is covered in the U.S. under 45 C.F.R. 46, Subpart B (Subpart A is the so-called Common Rule). The right to conduct research on fetal tissue is covered in the Public Health Service Act (42 U.S.C. 289 et seq.) as amended by the 1993 Revitalization Act. On the evolution of who should consent to fetal tissue research, see, e.g., Seifert 1997.

424(b)(3) of the Securities and Exchange Act 1933 summarized its "Research and Development" as follows:

The development of cells for therapeutic use will be an ongoing endeavor for many years and it is impossible to make any meaningful estimate of the nature and timing of costs related to these activities. . . . No specific completion dates have been established for any particular project since most of our work is experimental. We do not expect revenues from any R&D efforts directed toward cell based therapy for several years and these revenues may never develop if our research is not successful. We expect some revenues from research cells and media, but it is too early in our history to make meaningful predictions as to the amount of such revenues.

It then went on to specify 27 risks, some of them "Risks Related to Our Business" (although it made no mention of problems that might result from its egg procurement arrangements—see below) and some of them "Risks Related to the Securities Markets and Our Capital Structure." The second of the 27 listed risks was "We have a history of operating losses, do not expect to be profitable in the near future and our independent registered public accounting firm has expressed doubt as to our ability to continue as a going concern." How, then, did the company make ends meet while waiting for this highly uncertain research to pay off in an unknown amount of time? It applied for grants and public funding, and organized rounds of venture capital and private investment opportunities, all based on the promise of its central technologies. It prepared cell lines and media for sale. And it also pursued a sideline that was more immediately revenue-generating: cosmetics made with stem cells. In 2012, when new breakthroughs in R&D were being advanced, a face cream targeted at older women was the main product the company was producing that was making money. Its SEC prospectus also said this: "We are considered a development stage company and as such our revenues are limited and not predictable. . . . [skin care] accounted for $547,000 or approximately 51% of total revenue during the three months ended March 31, 2012."

The tag line on the company's shopping webpage beautifully summed up the company's view of the relations between skin care and more fundamental biomedical breakthroughs based on pluripotent stem cells:

Someday, stem cells will change the world.
Today, stem cells will change your skin.[11]

This kind of relationship between arguably rather frivolous though lucrative markets supporting more fundamental R&D may be common in the private sector, and may make perfect sense for the companies concerned, providing a safe income stream to subsidize the underlying research and protecting investors' interests.[12] The taxpayers of California who voted for Proposition 71, however, had reason to be concerned with how companies took up basic research done in the state with CIRM and non-CIRM funding. Skin-care products are not usually cures. I found in talking to students and others that it made a big difference to how many of them felt about funding research and donating eggs and other cells to research if the research to which donation was being made was biomedical in a strict sense. Most people felt more positively about donating to an open-ended, basic research process, as long as it was demonstrably on a trajectory "from bench to bedside," rather than primarily benefiting cosmetic, commercial, or military interests, even if there were trickle-down effects of economic activity or national security from which society might benefit from the latter.[13]

One response I encountered among young women students to the development of cosmetics was to argue that in that case egg donors ought not to be altruistic but to be paid a share of revenues; subverting the outcome of the research precipitated a change in motivation for donation. For certain women with whom I talked, the fact that stem cell companies were beginning to make money in cosmetic surgery or selling anti-aging creams to prevent women's skin from showing its age, suggested a system in which women were subsidizing the industry over the life course, through providing eggs for research when young, through paying taxes for the pro-cures promise that might not be accessible to them or their families because of price and/or which might never materialize during their middle years, and through being targeted by expensive cosmetics marketed to portray as flawed an older women's skin after childrearing.[14] Older men (enticed on Fathers' Day 2012 to improve their skin at a cost of $297 for 2 ounces with a dozen free golf balls) and some younger women were featured in product endorsements, but the majority of their customers appeared to be women in the over-45 category. For the science-society contract implicit in Proposition 71, the voters were to pay taxes to fund research for cures, and the possibility or necessity of industry partners getting off track doesn't fit neatly into this promise.

Cosmetics or cures (or hopefully both), procurement of women's eggs still had to be worked out. In 2010, the company began procuring women's eggs from two area fertility clinics under unspecified arrangements.[15] The financial model to cover the costs of egg procurement from patients and the clinics to the company was not made publicly available by either of the clinics or the company. In other parts of the world, treatment is subsidized if eggs are shared with researchers, and this may or may not have been the model here.[16] Under California state law, SB 1260, compensating donors for more than direct costs for eggs donated for stem cell research (whether or not funded by CIRM) was illegal in the state, at least until AB 926 might have come into effect. On July 24, 2013, CIRM held a meeting to consider whether or not CIRM funds could be awarded for work with stem cell lines made out of state without the same restrictions on egg donor compensation, and no doubt there will be pressure to bring CIRM regulations in line with state law, if SB 1260 is repealed. As far as I know, company payments to fertility clinics as intermediary brokers were not covered under either state law. Whether or not there is any financial arrangement, according to my reading of Proposition 71, the clinics could receive "reasonable payment" for cells for "the removal, processing, disposal, preservation, quality control, storage, transplantation, or implantation or legal transaction or other administrative costs associated with these medical procedures." The clinics could then choose to pass on some savings to women providing a share of their eggs in the course of treatment, as long as the company did not itself pay the women for their eggs. A change in the state law and then in CIRM's own regulations to match the new state law would likely be much more acceptable.

With the November 2004 passage of Proposition 71, the California Research and Cures Act, pluripotent stem cell research in California began to benefit from two kinds of public donations. One was state taxpayer funding; the other was donated tissue and cells.[17] Donations of human tissue and public funds for research were linked in the context of a voter-approved directed biomedical regenerative science initiative. Who was giving, what was being taken, and who was receiving these things, and what forms of reciprocity were (or should have been) involved? I turn first to public funding, and then to tissue donation.

For the first of the two public gifts, public funding, the act was written to "authorize an average of $295 million per year in bonds over a 10-year period to fund stem cell research and dedicated facilities for scientists at California's universities and other advanced medical research facilities throughout the state." According to the California state government's Legislative Analyst's Office, the fiscal impact of Proposition 71 was going to be as follows: "State cost of about $6 billion over 30 years to pay off both the principal ($3 billion) and interest ($3 billion) on the bonds. State payments averaging about $200 million per year."[18]

Federal funding for basic and applied research, and tax advantages for private R&D—and the public subsidizing of the cost and risks of research that this implies—had been vital to private and public scientific research in the United States and California for decades.[19] It was also business as usual in that, despite public funding of research, future profits could be privatized.[20] What was unusual was the amount of money coming from the state tax coffers, and the fact that the money was voted in "directly" through the California initiative process, by which voters earmark funds for specific goals or projects through amassing enough signatures to add the issue to the ballot. To some people, the California proposition process is inherently democratic; to others it is the exact opposite[21]; for my purposes here, the use of the proposition process to fund stem cell research in the state raised anew one of the issues at the core of science policy, namely, what the relations between public supporters and the funded research should be.

I found that there was a somewhat widely shared understanding of the underlying economic logic of the initiative for which the people of California had voted. Proposition 71 passed after more than fifty years of federal funding for science, and science policy that framed public investment in science as of importance to the nation.[22] The model of reciprocity implied in this federal model was primarily trickle-down, with open-ended (though potentially substantial) net gains to the economy along the lines spelled out in the proposition and quoted above. The model also included open-ended and ongoing gains to science and medicine, resulting from investing in ingenuity and innovation. Several people in different domains explained to me that public funding for the research was appropriate because, in theory, if profits to be made from commercializing any aspect of stem cell research returned to the researchers themselves and their institutions, innovation would be incentivized and subsidized, and the California economy would benefit through economic activity and tax trickle-down. Cures and innovation were the public goods on offer, and privatized profit was the carrot that kept the best and the brightest in the sciences, and the state in the black, ushering in the bio–silicon innovation economy. A strong statement of the role of government subsidies to overcome "market failures" in the bioeconomy was presented in the Obama administration's 2012 *National Bioeconomy Blueprint*:

a major justification for government investments in science and technology is to overcome market failures; these occur when private investors invest less in technology than the socially optimal level because they cannot reap the full benefits of their investment. In this context, scientific discovery is a public good that benefits all.[23]

In the stem cell case, the earmarking through the initiative process of funds for a particular field of research, and the promise of particular research ends (cures and a regenerated economy) in a very tight innovation cycle, presented a significant time challenge in the midst of ethical controversy and a worsening economy. A stem cell entrepreneur described venture capital's time span as being on the order of five years "in and out." CIRM's own ten-year funding cycle ramped up the time pressure, making the promise that much harder to fulfill.[24] There was a small window in which to vindicate the field ethically, scientifically and economically, and the underlying economic model relied on, at least rhetorically, (substantial progress toward) cures that might or might not realistically materialize in the time frame. This set up the potential for over-promising, a kind of innovation-logic driven statewide therapeutic misconception.[25] In addition, the fact that the public was footing an enormous bill in state taxes during a recession, in addition to federal tax spending on biomedical research, made several of my interlocutors, and CIRM itself, all the more attentive to the need for cures to be moved to the clinic as quickly as possible, and to be affordable to all the people of California.

Stem cell scientists I interviewed or interacted with told me that pluripotent stem cell research, in combination with related fields of research, would turn out to be revolutionary for science and medicine. To a person, however, they also articulated disciplinarily differentiated likely outcomes that were not the same as the implicit promise of cures in the near future that seemed to lie in Proposition 71 itself. One molecular and cell biology researcher put it this way: "Our lab is working on fundamental science and no one can really predict what impact it will have in the years to come. . . . One of us might get a Nobel prize or it might just be another brick in the wall." A bioengineer told me his lab was working on models that used biocompatible scaffolding populated with stem cells for tissue transplants that were "most likely going to be part of a fundamental toolkit" that would include 4-D modeling and advances in visualization. A clinical researcher told me that facilities with experience with hematopoietic stem cell transplants would be well placed to have significant pluripotent stem cell transplant clinical success, now that they were isolating pluripotent stem cells in tissue types with which they already have experience. An industry scientist told me that provision to scientists of reliably characterized and screened stem cell lines derived from embryos that were deselected during IVF for containing a disease mutation ("diseases in a dish"), and toxicity testing for drug development were where they were anticipating the highest payoff. The scientists expressing these perspectives were all bullish on the technology and its potential to revolutionize biomedicine, yet none of them—even though two of the four were working on specific diseases—expressed this excitement as being immediately in the form of *cures*. Translation takes time and can be unpredictable in its course.

Would it have been possible to acquire voters' assent to supporting this research without implying that cures would shortly follow? Does it matter that what the public voted for and what they were likely to get were rather different (after all, in the long term, the payoff for science and society might be greater than anticipated, precisely because the end products support such a wide base of procedures and knowledge)?[26] CIRM itself continued in their outreach materials to emphasize the promise of cures as evidenced by proof-of-principle animal studies, Phase I clinical trials, and grants to industry, and continued to pledge monies to this translational pipeline. In their more detailed strategic plans, however, they also set out and tracked goals that represented incremental gains in basic techniques and tool and model development, and in funding partnerships and educational efforts.[27]

The first public gift, then, was funding. The second ongoing gift to stem cell research was to be the donation of biological specimens (embryos, eggs, and other cells and tissue) as raw materials to be used in research. Once ethical protocols were established, researchers funded by CIRM grants began to be able to receive appropriately consented donations of human embryos, gametes, tissue, and cells within the terms of the state constitutional amendment.[28] The act established the right to work with human tissue:

There is hereby established a right to conduct stem cell research which includes research involving adult stem cells, cord blood stem cells, pluripotent stem cells, and/or progenitor cells.

It was implied that these tissues were to come from "research donors." In Article 3, "Definitions," one learned that

"Research donor" means a human who donates biological materials for research purposes after full disclosure and consent.

As was demonstrated in chapter 2 for California, in the United States the process of institutionalizing human pluripotent stem cell research brought a lot of attention to the question of whether at all—and if so under what conditions—embryos could be procured for research.[29] The procurement of women's eggs, while receiving less attention than many feminists thought appropriate, also received considerable policy attention, especially around whether and how much women could be compensated for donating eggs.[30] This attention fell off somewhat with the establishment of efficient ways of creating pluripotent stem cell lines from somatic (non-germ line) cells, so-called induced pluripotent stem cell lines. In talking to CIRM employees and grantees, it seemed that CIRM moved to reduce the "regulatory burden" of reviewing the procurement of somatic cells. They maintained that procuring human somatic cells—through skin punch biopsies or from medical waste—was not was not unique to stem cell research, and so high levels of attention from bespoke ethics committees for stem cell research were not thought necessary for *in vitro* iPSC stem cell research.

California is central to a recent U.S. legal history of granting property rights to human tissue to researchers and clinical and research institutions, while sedulously removing all property rights to bodily material and information from the person or group from whom the tissue was taken.[31] During the period covered in this book, however, many scholars and activists in California and elsewhere, myself included, were engaged in analyses of the ways in which the giving and taking of biological samples for research was increasingly out of sync with the scientific demands of the day and with the ethical instruments (like informed-consent forms) organizing the terms of the gift.[32]

A spate of high-profile controversies surrounding donation of biological tissue in the United States occurred during this period. These cases provided an environment in which questioning of the status quo on donation of tissue, gametes, and embryos for stem cell research resonated. They included a new wave of scholarly and popular attention to Henrietta Lacks, the African American woman whose cervical cancer cells gave rise to the prolific HeLa cell line, and drew out the themes of race, class, and gender and their relations to the new tools and commerce of biologicals, access to medical care in America, and a troubled history of informed consent.[33] Two decades of the use of Havasupai blood samples procured in the context of diabetes research by Arizona State University, and subsequently used for unconsented research, including research that the tribe found objectionable (whether or not their samples were anonymized) on migratory origins, in-breeding, and schizophrenia, led to a lawsuit in 2008 and a settlement in 2010.[34] The storage and use for subsequent unconsented research of infants' blood taken via a heel prick during routine new-born screening was contested in Texas, where giving the samples to the federal government particularly rankled, and Minnesota, where the incompatibility with the state's genetic privacy law drove the case, resulting in decisions in 2009 and 2011 respectively.[35] The threats to one's own and one's family and community privacy from ever-expanding forensic DNA databases was often in the news, as was the fact that these databases were disproportionately sampling some communities.[36] And in California two prominent universities hit the news as they solicited and tested their students' DNA.[37] A keenly watched battle over breast-cancer-gene patents reminded the public, especially women, that whether or not a breast-cancer gene is "a product of nature" is part of deciding whether the knowledge and techniques based on them are a commons of sorts, or whether individual companies can control and profit from research and diagnosis.[38] Within U.S. pluripotent stem cell research itself, questions were raised as to whether some of Harvard's stem cell lines could be part of the 2009 expanded NIH Stem Cell Registry because they were consented in the context of diabetes research and consent to do different kinds of research with the cell lines could not be presumed.[39] In a more neoliberal vein, threats to the privacy and confidentiality of donors,[40] and various institutional bio-bank-initiated models for participation in research by donors and communities,[41] also garnered a lot of attention in this period.

In many ways, then, the timing was right to rethink the question of tissue donation to research in the state. It is often thought that public understanding of science is important to avoid misunderstandings and to secure support for science and for its role in governance.[42] In recent years, especially in Europe, there had been a great deal of enthusiasm for public participation in science governance and decision making. The campaign for Proposition 71, and CIRM after its passage, engaged in various kinds of outreach and educational events and initiatives, especially in conjunction with disease advocates, to educate and convince the electorate to fund the research and to continue to support it. Efforts to improve the public understanding of science, however, were not the only way in which relations between the public and the science needed tending, given that the research benefited not just from money but human tissue as a substrate. As up-front donors, the public was implicated in the research not just downstream, but upstream too, before there were therapies or other profitable products, and before a lot of the science was worked out. In what ways did the two kinds of interconnected donation obligate and entitle the public donors?

As table 5.1 shows, there were *prima facie* difficulties in reconsidering how to organize the donation of human tissue for pluripotent stem cell research, stemming in part from the variety of human tissue involved. There had been calls for a general framework for the donation of different types of human tissue, especially in Europe. For example, the so-called European Tissue Directive (Directive 2004/23/EC Of the European Parliament and of the Council of 31 March 2004, enacted 2006–07) expressed the need for a unified framework as follows:

The availability of human tissues and cells used for therapeutic purposes is dependent on Community citizens who are prepared to donate them. . . . There is an urgent need for a unified framework in order to ensure high standards of quality and safety with respect to the procurement, testing, processing, storage and distribution of tissues and cells across the Community and to facilitate exchanges thereof for patients receiving this type of therapy each year.[43]

Despite its call for a unified framework for human tissues and cells, however, this directive in fact covered "haematopoietic peripheral blood, umbilical-cord (blood) and bone-marrow stem cells, reproductive cells (eggs, sperm), foetal tissues and cells and adult and embryonic stem cells," but it excluded "blood and blood products," "human organs," "organs, tissues, or cells of animal origin," and "tissues and cells used as an autologous graft within the same surgical procedure." It also applied to donation of tissues and cells "intended for human application," but did not apply to research using human tissues and cells for "in vitro research or in animal models." Because of its focus on cells and tissue for "human application," safety, accreditation, quality, traceability, and infectious disease screening were its primary concerns. Under the slogan "We are all potential donors," donation was urged as a matter of public health and safety and of community self-sufficiency. Consent to donation was covered in Article 13, but the only requirement was that national standards of consent to donation be met. In terms of how to feel about and orient oneself ethically toward donating, the directive called for "voluntary and unpaid donation, anonymity of both donor and recipient, altruism of the donor and solidarity between donor and recipient." The scope of the Tissue Directive did not shed much light on tissue donation for pluripotent stem cell research, then, because of all these differences.

Policies and practices relating to tissue donation for stem cell research varied greatly depending on the type of body tissue being procured, whether embryos, eggs, placental cells, cord blood, fetal tissue, foreskin, or other skin.[44] The frame of donation—whether altruistic, a compensated gift, a diversion of medical waste, a diversion of reproductive potential, or a diversion of some other treatment, for example—varied. The legislation relevant to these different frames, and the legislation's aims also varied. Depending on the tissue sought, and the setting of its procurement, donor protection or consent, abortion relevance, environmental and public health and safety, infectious disease control, or tissue matching and allocation were the major rationales behind legislation. And behind all these differences, even just within California, were long, multiple, culturally specific, religiously and/or civically meaningful trajectories for the treatment of body parts.

In view of these difficulties in rethinking donation and reciprocity for stem cell research *ab initio*, a more embedded approach seemed called for. Ethnographic observation of sites and occasions where donors, donated tissue, recipients, and their funders and regulators interacted and contested reciprocity provided insight into the ethical choreography at work. The ambient ethical and regulatory documents and academic models proposing different solutions to research donation worked with ethnographic observation as a triage system out of which to follow certain tropes, and potentially begin to transform, tissue donation for public-private stem cell research.

Who were potential donors to human pluripotent stem cell research? Sometimes a donor presented as an individual who was giving cells or tissue or bio-information, and had sole decision rights over her donation. This fitted pretty closely with the imagined signatory of informed-consent forms. Scott (a pseudonym) was an undergraduate at a top public university in California. He told me that he was going to get his DNA "read and banked" with a genomics company once he had a job to pay for it, and that he would, "no problem" sign up for tissue donation for genetic or stem cell research. He didn't have a problem with privacy, and he wanted to help science. As he talked on, it became clear that his willingness to donate was not simply an uncomplicated individualism. He was also rejecting, quite forcibly, certain aspects of what he saw as general academic tendencies of a leftist university (which he liked in most regards) to be insufficiently libertarian. He was both attracted to and annoyed by certain faculty members and certain approaches to scholarship in the social sciences and the humanities; these classes had truly changed his way of seeing the world and that was why he defended taking them to his friends in the sciences and engineering who thought he was "insane" to spend so much time this side of campus. But he also thought that these professors and classes were often too knee-jerk critical and negative, and too behind the times to understand or change a world in which things like global big data and regenerative medicine were just a part of reality.

In regard to donating tissue for stem cell research in particular, he couldn't understand all the molly-coddling around returning medically relevant results or incidental findings to donors through doctors or other "experts," wishing that everyone would "just grow up" and use the information to good effect. He was irritated by the fuss about commercialization of human tissue, especially complaints that it was unjust that only recipients and not donors could have property interests in cells and their products: "If you want a cut, do the science, start the company—it's open to anyone to do. . . . And if you can't then maybe you'll appreciate why . . . we reward that in this country!" At other times, I had heard him critique excessive commercialization, and argue for community-based benefit sharing, so I knew that these were not his only thoughts on the subject. But his position made clear that feeling that he fit rather well the implied individualism of the donor who signs an informed-consent form did not mean that he took his individualism for granted. Rather, the autonomous individual he could approximate in a forward-looking state like California was something to be fought for as much as submitted to or taken for granted.

Tiffany, Adela, and Matt (also pseudonyms) were graduating seniors at the same public university, and all three of them were cautiously willing to be donors to stem cell research. Unlike Scott, however, they saw themselves from the get-go not just as individual donors, but also, perhaps primarily, as members of groups who would be implicated by their donations. In Tiffany's case, she was interested in possibly donating some of her eggs to stem cell research. She was aware of the risks of egg donation, forgone income (from not donating to fertility clinics), and wider social meanings of recruiting as egg donors young women, especially young, bright, slim Asian American women like her. Adela, on the other hand, was sure she did not want to donate eggs, but she was interested in donating skin or other somatic tissue to stem cell research, so as to add diversity to the local bio-bank and to science. Difficulties with access to health care and disparities in health were lived realities and were crucial to how she thought of herself as a donor. Matt was interested in donating so as to help his brother who had a severe cognitive disability that his low-income family had few resources to address. He thought that, given their genetic closeness, stem cells from him might either be good candidates for transplant therapy for his brother, or be useful for drug testing or pre-clinical studies of some kind. At no point did he describe himself in relation to donation as anything other than tied into this close kinship.

In other words, ethnographically, a donor of cells, tissue, or bio-information to regenerative medicine could be positioned vis-à-vis their donation as an individual, and /or as someone with group membership (as a "woman" perhaps, if donating eggs; or a disease advocate, perhaps, with affected cells that could yield stem cells from which to study the condition or its treatment *in vitro*; or as a member of a demographic group, or of a family, whose communities or family members might be involuntarily implicated). This implied that donation meant different things and needed to be able to reflect that in various ways.

The recipients of donation also were ethnographically varied. As recounted above, scientists in different disciplines reported different goals of their research. Those same four scientists also related differently to donors. The clinical researcher did not expect to start a company or make money from products that began as donations. He was extremely experienced with and thoughtful about the rights of both donors of tissue and patient recipients of donated cells. Being concerned with graft versus host disease and histocompatibility, he was also acutely attuned to community and family aspects of donation to research and therapy. The industry scientist wanted to avoid ethical and regulatory roadblocks, and chose his non-embryonic tissue source on that basis. His company also wanted to be able to scale up, so the ready availability of his source tissue was important. Similarly, quality control and therapeutic standards were important all the way up the product chain, from original tissue donations through to reagents and characterized cell lines, so documentation and post-donation curation of the gifted cells were also essential.

The molecular and cell biologist preferred to keep herself at one remove from donation by procuring her cell lines from acceptable biobanks or from trusted clinical colleagues. She was somewhat troubled by the way some of her colleagues took ethical procedures less seriously than she thought they should, and a bit put off by the rush by some to jump into commercial ventures. She pointed out, however, that in a time of budget crisis, universities were actively rewarding faculty members' commercial activities with promotions and other benefits. She didn't think it was a serious problem but she did think that certain trends worked to make some recipients of cell lines less thoughtful than they should be about donors' rights and responsibilities. The bioengineer was working on the biocompatibility of his scaffolds and was happy to use the least problematic, most readily available "acceptably derived" stem cell lines for the moment. For different reasons, then, the clinical researcher and industry scientist took the organization of research donation very seriously. The bioengineer and molecular and cell biologist kept some distance from donation, though especially the latter cared a great deal about deeper questions about the donation of human tissue to stem cell research.

As ever in this field, what people like donors and scientists said about themselves and what they did interacted with the documents and regulatory, ethical, and promotional acts, instruments, and occasions. The latter that were part of the field, but they also guided or critiqued it. CIRM's 2012 Strategic Plan included many lists of the recipients of its funding and scope of its operations, as it envisaged different aspects of its mandate. The most commonly named recipients were "grantees, industry, other government agencies, disease foundations, venture capitalists and others," and "academic institutions." The "broad community" was mentioned, but in the context of needing to be kept informed and involved. Recipients, like the potential donors listed above, did not come to research donation from an abstract position, even though they were all organized by, and often contributed to, ethical protocols that strove to articulate meaningful guiding principles and uphold binding—if evolving—ethical standards.

And then there was the biomaterial itself, which was billed as pluripotent and immortal, with traces and histories of its own that were slowly becoming clearer in their implications. Debates regarding the right way to organize relationships between these three actors (donors, recipients, biomaterial) took on salience at particular points during my research. The following were summaries I made while analyzing my data of the times that the elements of these relationships were actively debated during my research. They contain themes that echoed in the literature at the time, so I have included the footnotes I added to them as I carried out my research:

• Discussion of what the relationship between the donor and the biomaterial should be arose in debates about egg donation and in that context prominently included compensation, and to a lesser extent, affect (how to feel).[45] The relationship between donor and biomaterial came under scrutiny less frequently in discussing the donation of somatic tissue, but when it did come up, these debates tended to focus on informed consent, and on rationales for keeping the donor and the biomaterial connected: ways that the donor might benefit or be at risk from (herself or others) knowing things about or having access to the stored products of the biomaterial, and ways that the value-added products from the initial donation might be enhanced by learning more about the donor over time.[46]

• Discussion of what the relationship between donated biomaterial and recipients should be occurred in debates about access to stem cell treatments, especially publicly funded treatments, and in debates about histocompatibility, where stem cell lines from one's own cells, or family or community donation, were seen as technical and social goals to be pursued.[47] Unlike for organ, milk, or blood donation, principles of triage (who to treat) of recipients in the face of scarce donations or expensive treatment did not ever come up during my research, perhaps because treatments were still in the future. The relationship was also implicitly at stake in the other direction, from recipient to biomaterials, in the organization of patient advocates and the efforts of scientists to give biomaterials their greatest potential to lead to efficacious products.[48]

• Discussion of what the relationship between recipients and donors should be arose when patients and when scientists were the recipients. For example, donors to patients were encouraged to feel the appropriate altruistic motivation for donation by empathizing with those needing cures without expecting oneself or one's loved ones directly to benefit from the donation. Altruism, whether through the more European "solidarity" or the more American "pro-cures" logic, contained within it the idea that any one of us, or our loved ones, could need the donation at any time, without needing a more explicit kind of reciprocity.[49] There was also some concern that patient recipients should not, as a class, be more privileged than donors, and thus risk exploiting donors as a class. As far as donation to research was concerned, scientists and research institutions desired, and bioethical instruments were developed to implement, donation based on the same altruistic affect on the part of donors. In so far as they were working in the public interest, scientists and public and private research institutions and companies were seen as vital intermediaries between donor and recipient making cures possible; as unique possessors of innovative intellectual capacity they were thus covered by the altruism logic of donation to patients. The level of public subsidy and the rise of commercial activity by research recipients in California, not all of it related to cures, however, strained this view, and re-opened the question of the appropriate relationship between researchers and donors. Some argued that altruism on the part of donors was not necessarily the best affect to guide the different parts of the relationship between donors and stem cell scientists, research institutions, and companies as recipients.[50]

In short, donors, recipients, biomaterial, and the contexts in which the relations between each were at stake, had to be approached in an embedded way. Were there, nonetheless, models to draw on to address the strains in the relations between donors, recipients, and biomaterials arising during the process of normalizing stem cell research in the state?

RECIPROCITY WORTH FIGHTING FOR?

EGGS AND COMPENSATION: A COMPARISON WITH COMMERCIAL EGG DONATION TO ARTS (ASSISTED REPRODUCTIVE TECHNOLOGIES) IN CALIFORNIA

Proposition 71 made it legal to procure eggs for stem cell research, igniting debate and inviting comparison with egg donation for assisted reproductive technologies in the state. Under the terms of Proposition 71 egg and embryo donors were required altruistically to donate their tissue:

Medical and Scientific Accountability Standards b) The ICOC shall establish standards as follows: . . .

(3) Prohibition on Compensation

There was widespread agreement that the embryos surplus to IVF treatment that could legally be donated to CIRM-funded stem cell research after the passage of Proposition 71 should be donated without compensation. The question of egg donation, however, posed a greater challenge, because commercial egg donation for fertility treatments was thriving in the state at the time. If egg donors who donated under one set of conditions could be remunerated but not under the other, why would anyone donate to stem cell research rather than to IVF? And if it was fair to pay women who donated eggs for fertility treatment, was it not unfair not to compensate them for their invaluable role in stem cell research? Could commercial egg donation in California be a model for egg donation for stem cell research in the state? The comparison—and, the competition— between the two kinds of donation, and resultant scarcity of uncompensated donated eggs for research made sure that the question of egg-donor compensation was addressed. Arguments in favor of compensating egg donors for research fell into three categories: ways safely to incentivize donors for the risky and time-consuming process (promulgated by scientists and clinicians and other supporters of stem cell research to overcome shortages), ways to avoid creating an underground market or exploiting a small class of women willing to donate eggs without compensation, and irritation that egg donors who would be compensated for fertility treatments should be asked to forgo compensation in the case of stem cell research, whereas no one else was being asked to forgo their possible monetary stakes in the research (quite to the contrary, the innovation cycle depended on it).

The question of how to circulate *ex vivo* human embryos and gametes in assisted procreation had been debated and practiced for more than thirty years. Work on appropriate payment levels to avoid undue inducement, however controversial, had been carried out.[51] Effort had gone into how to make, break, and put on hold chains of custody that would determine who was a parent when there were several contenders all contributing to a pregnancy, and to which parents and which state a child born with ARTs belonged. A consensus of sorts had arisen around the concept of the 'intended parents', according to which donors of gametes and embryos to reproductive technologies needed to take themselves out of contention for parenthood to any children conceived from their donation. The following source for how a California egg donor should ideally behave is consistent with fieldwork I have done on the topic in the past.

A reputable California egg-donor brokerage explains what is to be expected during egg donation this way: The agency's role is to "connect outstanding young women who are willing and able to give this life-changing gift to recipients looking to build a family."[52] The recipient of the eggs (or her surrogate) will receive eggs from someone she has chosen from a psychologically and medically screened database of available donors whose pictures and information the recipients see. An academically stellar donor costs an extra $500. If both donor and recipient agree, they can talk on the phone anonymously or meet, but they do not have to. "The recipients assume all moral, legal, and financial responsibility for any children born from the egg donation. Egg donors have absolutely no parental rights or responsibility for the future welfare or support of these children." The recipient pays all costs, including the donor's attorney fees for making the contract. The recipient gets a baby, with ongoing obligations to that child (the future rights of the child to contact the donor are unspecified, and subject to change).[53] Recipients are due "care and consideration" because the "time and expense related to infertility, not to mention the tremendous emotional toll that it takes, means that a higher level of compassion is required." The donor also receives something when she makes her "life-changing gift," though it is not in exchange for the gift itself. For her "time, energy and commitment," she is "compensated twice: once with a financial payment and once with the sense of fulfillment" she receives "from helping to complete a family!" In case there is any doubt, these two kinds of reciprocity for her egg donation go hand in hand, with the money being given in the context of the already present correct affective orientation to the gift: "In addition to the personal satisfaction you will have knowing you have helped create a family, you will receive compensation for your inconvenience and time, and all of your expenses are paid by the recipients."[54] Recipients are equally reassured that donors will have been screened regarding "their motivation for becoming an egg donor," and are committed to seeing through the gift. The donor should display "care" rather than receive it.

To help prospective donors to know how to behave so as to be chosen, and in case it is hard for the donor to walk away from the donation once compensated, there is an "ask an egg donor" webpage where prospective donors can have their errant feelings and opinions realigned with the nature of the gift and its allocated reciprocity by a star former donor.[55] Blond-haired and blue-eyed in her accompanying photo, she draws in an upbeat manner on her own experiences. She presents herself as a donor who had no lasting side effects, and who is completely happy with her decision to have been a repeat egg donor. She is not quite as romantic about egg donation as the clinic's web page, but she insists on knowledge, consent, good contracts, and willingness to be alienated from your donation. She responded as follows to someone who worried about her eggs, or embryos made with her eggs being diverted to stem cell research:

I was asked to sign a waiver (that I was not, in any way, pressured to sign) that said stem cell research may or may not be something that happens with my eggs once they're retrieved. If you're not on board with that, then you needn't sign the release. . . . Egg donation is a business; it's a business that gets ALL up in your business. And once you sign your consents and receive your pay, the eggs—and any child—that result are not your business, to be frank.

When she was a donor, she tells us, she made sure she was protected by an air-tight contract, disclosed her donation to the right people, and felt no need for further contact with the recipient or any resulting children about whom she had no doubt that they belonged to the recipients. Future children's rights to contact the donor and complicate the picture of a one-off gift and reciprocity are not mentioned.

All the gift-reciprocity affective scripts for recipient, donor, and even to some extent, the broker and other professionals involved in the transaction, are rehearsed on the web and throughout the process so that each player can learn how to behave, and thus how to give and receive appropriately relative to one's role. Compensation was perceived as rewarding the donation and effecting the necessary alienation between gamete and kinship roles. The eugenic aspects of the pay structure of third-party ARTs discussed in chapter 3, and the incentives that payment for egg donation provides to discount the threats to women's health of egg donation mean that it is problematic to pay women for eggs under any circumstances. But payment makes a certain sense. It classifies the form of reciprocity in the exchange as one that is not wholly based on kinship, care, or other kinds of ongoing connection, rights, or obligation. In short, despite all the problems with paying egg donors, payment reflects the significance of the gift, and the necessary rupture of your connection to the child. The money paid is not a counterargument to donating "for the right reasons," but the opposite—paying for egg donation is one means of valuing a donor for her gift in a currency that does not establish ongoing connection; in theory, the donor has neither rights nor obligations to the future child(ren).

In the worldwide egg-donation industry, the problems cited above about eugenics, undue incentives, and health risks, were serious and still had not been tackled adequately. As reproductive labor for others becomes more and more normalized around the world between rich and poor both within and across borders, and as it becomes an increasingly routine kind of gendered, classed, racialized labor keeping families financially afloat in hard times, dealing with distributive and economic justice aspects of third-party ARTs remained as urgent as ever.[56] It is also not clear that the egg donor really can or should disconnect completely from her donation. Calls to collect health-care data and socioeconomic data on donors for their future protection need donors to be accounted for and tracked over time. And a global groundswell of activism to move away from donor anonymity, spearheaded in large part by donor-conceived children, was already underway, making it more and more likely that egg donors would be contacted in the future by children conceived from their gametes, regardless of how well they choreographed their alienation from their gift.

Could egg donation for stem cell research in California have been built on this model in terms of the implication of the donor in the research as measured by the form of reciprocity? In some ways it is possible to analogize the gift of egg donation for stem research to the gift of life in reproductive technologies, and the star egg donor's response about stem cell research above suggests that the two are not in any case firewalled one from the other. (The egg-donation agreement of the stem cell company discussed above also suggested as much.) There are important differences: the proximate recipients of egg donation to stem cell research are scientists and their companies and institutions and not those for whom the research might ultimately be a gift of life; the eggs will have been turned into something very different before they reach patients, and thus the form of help being given in stem cell research is far more nebulous. There are significant parallels, too: the first parallel is the identical procedure and its significant—remuneration worthy?—risk level. There is also a parallel in that the recipients in both cases (the intended parents for assisted reproductive technology donation, and scientists and their institutions for stem cell research) often want the donor to alienate or detach from her donation.[57] Intended parents want it to be clear who are the parents, while scientists have to be free to carry out research on donated cells, and in California must also exclude the donor from any property interest in the donated tissue to follow the incentive to innovate argument rehearsed above. Oddly, another similarity lies in the fact that there are limits to the degree of alienation that is possible in both cases. Egg donors to fertility treatments are likely to be contacted by donor-conceived children down the line, however well they took the star donor's advice, and detached (until the next time) with money, excellent legal representation and medical care, and the glow of the gift. For their part, egg donors for research are likely to be asked to stay in contact with the scientist recipients for the purposes of corroborating health-relevant information over time (see below), so there are also pressures that work against the desire on the part of recipients that donors detach from their gift. The problem with the parallel surfaces, however, once one begins to examine the relation between the desirability of disconnection and the form of reciprocity that underscores that. In commercial egg donation for ARTs, reciprocity takes the form of a monetary compensation, plus feeling good. Money is, as forms of reciprocity go, good at ending a transaction. In egg donation for stem cell research, egg donors may not be paid beyond reimbursement for direct expenses, and their feeling good is focused on a more diffuse target. This comes close to providing no meaningful reciprocity at all. It is hard to imagine what "Ask an Egg Donor" could recommend to someone who was donating for research. It seems likely that she would recommend that prospective egg donors donate to ARTs rather than research, for the same kinds of reasons that she recommends that prospective ART egg donors switch from less to more reputable egg-donation outfits: the protections and benefits and rewards, for the right kind of person, are just much better.

The advent of human embryonic stem cell research using women's eggs was accompanied by regulation and rhetoric aimed at shoring up the altruism of donors to research through the banning of payment to donors. Progressives and women's groups wanted to protect women from the medical risks involved and thus wanted to disincentivize donation. The framers of Proposition 71, to avoid controversy, and because the funding was public, were very careful not to commercialize egg donation or in any way appear to be buying and selling body parts, let alone potential reproductive tissue. The model from commercial egg donation for ARTs, despite several structural similarities and an identical technical process to egg donation for research, did not in the end help much with working out the appropriate kind of reciprocity to reflect donor implication in research. Barriers to paying members of the public, as opposed to private egg donors for IVF, for egg donation were simply too high, and without payment, the productive part of the parallel—the need for a form of reciprocity that allowed provisional disconnection—was lost.

HUMAN SOMATIC TISSUE RESEARCH IN THE AGE OF BIO-BANKING

Once induced pluripotent stem cell research took off, pluripotent stem cell research was able to rely less on eggs and embryos for generating new stem cell lines. This promised to alleviate some of the problems with procurement and reciprocity that surrounded eggs and embryos, but it opened up the question as to how this public initiative was going to procure somatic cells for the new kinds of pluripotency, and with what forms of reciprocity. It was a time of an explosion of innovation in genomics and tissue banking, and there were urgent calls for revisions to informed-consent protocols covering the donation of tissue to biobanks and to research. There were several aspects of informed-consent forms that were no longer considered adequate.[58]

In April 2012, CIRM sought comments on its March 2012 Draft iPSC (induced pluripotent stem cell) Consent for Ongoing Donor Interaction.[59] This model consent form template was not meant to be the last word on donation ethics. It was meant to reduce bureaucratic redundancy, and it was designed to be compatible with California and CIRM regulations, to resonate with national regulations and guidelines covering federally funded research, and to be consistent with the International Society for Stem Cell Research model consent forms. There were a cluster of related reasons to do with scientific developments in bioinformatics and regenerative medicine that put pressure on key elements of informed consent, as expressed in the California Experimental Research Subject's Bill of Rights, and elsewhere.[60] They included the following, which I have illustrated with excerpts from the CIRM iPSC model consent:

1. Confidentiality and privacy could no longer be guaranteed to research donors, because of the identifying DNA contained in tissue samples; in any case, researchers increasingly wanted to retain contact with donors. This was evident in the iPSC draft consent, which was titled to include "ongoing donor interaction."[61] It included the following language about what researchers would want from re-contact (donors would, in turn, be able to have incidental results returned to them should there be any, if they so consented):

In the future, we may want to contact you to obtain additional information regarding past or current health conditions. The information may be important for research, developing medical products or treating disease.

And it acknowledged the risk to privacy and confidentiality from the DNA in the specimen[62]:

Since iPSCs made from your cells contain your genetic information, there is a very small risk that they could potentially be linked to you in the future even though we will have removed all links between the cells and your identity.

The second element of informed consent was as follows:

2. Research donors could no longer withdraw consent at any time, given that once turned into iPSC lines, the tissue would proliferate, and specimens might be widely distributed, integrated into biological products, and not able to be recalled.

The iPSC model consent noted this problem with withdrawal of consent and spells out the options and limits of the principle:

> What are my options for withdrawing? At any time you may exercise your right to withdraw from this research project. Prior to creating iPSCs: Option 1: Withdraw use of donated blood, hair or skin. You may withdraw your consent for the use of donated blood, skin or hair until the process of creating iPSCs has begun. After creating iPSCs: Option 2: Request that your cells be made anonymous by removing all links to your identify [*sic*]. By removing these codes, researchers will not be able to contact you for any reason.

In other words, exercising one's "right to withdraw from this research project" "at any time" was only really possible before the creation of iPSCs. The third element was as follows:

3. Donors could not be informed about all the research for which their tissue might be used because there was no way to know ahead of time what kinds of research would be undertaken with banked cells and products derived from them in the future.

On this, the CIRM iPSC model consent merely says, under "Key points to consider before you consent to participate," "Future research unforeseen at this time." Tissue donors could not be informed about the future research to which they were consenting, so their consent could not be informed about that research, its merits, or its risks. They could, however, direct their donation, but scientists would be free not to use directed donations:

> You have the right to place additional restrictions on how your specimens are used. The research team may choose to only use cells from donors who agree to all future uses without restriction. . . . You may not place restrictions on who may be treated with your cells or resulting medical products.

In addition to these three ways in which the research had outgrown the fundamental principals of informed consent, there were also reasons for dis-ease that originated from justice-based perspectives:

4. Expecting donors to get nothing while expecting recipients to make money off donors' tissue, exacerbated by increasing desire of researchers to get other kinds of information and data from donors.

On this matter, the iPSC model consent simply reiterated the status quo:

If any medical products result from your participation, you will not be entitled to any of the profits associated with such products.
    You will not be paid for participation in this study. You are eligible for reimbursement for any direct expense incurred as a result in [*sic*] participation.

The fifth element was this:

5. The increasingly obvious parallels with biological resource extraction in the agricultural and pharmaceutical biotech sectors, in which a generation of experience with benefit sharing schemes already existed.

On this one, the model consent was mute. But commentators on bioprospecting and biobanking had considered it.

FOUR MODELS AND THEIR PROVOCATIONS

If pluripotent stem cell research in California at this time had been entirely privately funded, perhaps a framework for tissue donation could have been worked out in which the kinds of quasi-markets found in gamete donation for assisted reproductive technologies in California would have evolved. It was not plausible, however, for the state to operate a trade in human tissue procurement, however well reasoned the level of payment. Likewise, if the research had been set up to be entirely non-profit, with no one making money from it, perhaps an altruistic model would have worked without friction. The research effort was supposed to make cures and money, though, as part of its promise to the taxpayer. The presence of the profit motive further up the value chain made a difference. A rethinking of the status quo on human tissue donation—to deal with the five problems just enumerated, for public-private stem cell research in the state—was called for.

Four models were devised and/or readily available in California at the time that rethought the relations between donor and recipient and bio-material / bioinformation. They were "open consent," "propertization," "benefit sharing," and "in kind." Payment for cells over and above direct expenses, as discussed above in the section on egg donation, did not receive serious attention. The first of these, open consent, came out of genomics, at just the right time.

OPEN CONSENT

At the 2008 SciFoo Camp (an O'Reilly Media–Nature Publishing Group–Google organized invitation-only scientific "unconference" held at the Googleplex each August to which I happened to have been invited that year because of an interest on the part of the organizers in seeing whether social scientists had anything to add to the mix), I learned more about a recent proposal to reform research donor consent to an "open" model. I had first learned about this proposal the year before when a working draft discussing needed changes to informed consent had been circulated.[63] The working paper discussed the "comprehensive identifying" genetic informa-tion the researchers were seeking to calibrate with phenotypic traits and medical histories, and suggested that consent forms should make it clear that privacy cannot be guaranteed anymore. One should only participate if one were comfortable with having this level of data out in the open. It also proposed working only with highly educated donors, and being very open about the implications of findings for one's family and community. Problems enumerated above in the discussion of the model iPSC consent that were for the most part being ignored, downplayed, or papered over elsewhere were to be front and center of this open consent. "Rather than shying away from this difficult issue," the authors wrote, "the open consent protocol straightforwardly addresses it."

Several prominent genetic and genomic scientists were at the 2008 SciFoo, including Walter Gilbert and his collaborator on the first direct genomic sequencing method, George Church. Church had recently launched the Personal Genome Project (PGP, as opposed to the HGP, or Human Genome Project), for which the open-consent plans had been adopted.[64] The PGP aimed first to compile complete sequencing data on all 46 chromosomes of a group of prominent genomics enthusiasts, and then to open the project up to 100,000 eligible adult participants. The whole genome data was to be matched with participants' medical records and phenotypic and other measurement data and made (after an opportunity for redaction) freely available to the public on the Internet.[65] Among the original ten volunteers to the project (known as the PGP-10), George Church himself, and Esther Dyson were at this SciFoo. In view of the Google tie-in, it was not surprising that the founders of 23andMe, the most-talked-about direct-to-consumer personal genomics company in California, were also present.[66] James Heywood of PatientsLikeMe was also present. The personal genomics sessions were not exactly the interdisciplinary mind-melding I had naively imagined would be at SciFoo, and (Peter Pan syndrome and non-hierarchical "we're all geniuses; everyone is worth listening to" rhetoric notwithstanding) the sessions involved the principals doing largely uninterrupted show-and-tells to large, well-behaved audiences. There was, then, no heated debate about donors, scientists, markets, consent, and possible conflicting approaches to reciprocity in the regenerative and genomic era in the formal sessions. There was plenty of chat and excitement behind the scenes, however, and the whole meeting was a great opportunity to get a sense of the ethos and varieties of the "spitterati."[67]

PatientsLikeMe presented itself at the meeting as unabashedly for-profit, despite strongly espousing the "openness philosophy" where patients would give their data without pay and allow it to be open. 23andMe, which like other DTC genomics companies charged a hefty sum for its services (and where I have my DNA), quickly began crowd-sourcing its clients for self-reported data.[68] It applied for its first patent in 2010.[69] At the meeting in 2008, it already seemed likely that the open-consent model espoused by George Church (who was backed by Harvard) was not one that would readily fly for California companies like 23andMe or for state-sponsored research such as that funded by CIRM. Church and colleagues were offering a kind of quid pro quo of openness, and backing up their request for radical openness from screened and highly educated donors with various kinds of open-source gestures of their own regarding their software and findings. The California model, insofar as such a thing was evident, was not like that. Whether a private company or a publicly funded initiative,

California seemed to be comfortable with—indeed, to demand—an innovation model that incentivized scientists and funded R&D by privatizing intellectual property, imposing few requirements of openness on companies, but asking for altruism without reciprocity from donors.

In 2011, the U.S. National Academies produced a report of a committee chaired by Susan Desmond-Hellmann, M.D, M.P.H. (chancellor of UC San Francisco since 2009, and formerly of Genentech and Bristol-Myers Squibb, San Francisco's Economic Advisory Board, and the California Academy of Sciences). The report was titled *Toward Precision Medicine: Building a Knowledge Network for Biomedical Research and a New Taxonomy of Disease*.[70] Central to the National Academies' proposal for a new taxonomy was an "Information Commons and Knowledge Network." This infrastructural initiative, designed to be "a blend of top down and bottom up activity" (p. 2) integrating "basic biological knowledge with medical histories and health outcomes of individual patients." The report was written with the needs of clinical scientists and the biotech industry centrally in mind. It emphasized the importance of, but did not specify the particulars of relations with, donors of tissue and information beyond stating a need to move forward flexibly, allowing patients (and researchers) access to the data, and be open about the real risks to patient privacy. It was clear about the change that needed to occur in the innovation model to disincentivize proprietary databases, and the social changes needed for industry and patient buy-in:

Importantly, these standards should provide incentives that motivate data sharing over the establishment of proprietary databases for commercial intent. Resolving these impediments may require legislation and perhaps evolution in the public's expectations with regard to access and privacy of health-care data. (p. 6)

Since the publication of iPSC discoveries in 2007 and 2008, other open-consent models were being devised in the United States that would compliment the National Academies' 2011 report. By 2012, John Wilbanks' Portable Legal Consent platform (hosted by Sage Bionetworks' non-profit Synapse computational research environment), from his Consent to Research project, had received prestigious funding, endorsements in top press venues, and buy-in from some researchers and disease organizations and was ready to launch.[71] A two-year NHS study reported on returning findings to donors.[72] The stage had been set to begin to forge a new kind of alliance between donors and recipients that made data and knowledge access a right of the emerging big data architecture in personalized medicine. If these open databases were to outpace in quantity, integration, and quality any proprietary databases, private companies—even in California— would likely see the business sense in moving beyond proprietary databases, especially if there was resistance from donors demanding the right to access to data and knowledge. Work on stem cell bio-banks could follow suit, with the right "in kind" quid pro quo. (See below.)

PROPERTIZATION

Stem cell research in California had to overcome a perception among some that pressures to commercialize science, especially the biomedical and life sciences, were threatening the integrity and openness of science, and leading, counter-productively, to over-promise on the likely fruits of research. One response to these concerns, and to the fourth reason above for informed consent no longer working (the concern that it was not right that recipients but not donors could profit from donors' cells), was to propose that donors got or retained a property interest in their cells, despite the legacy of *Moore v. UC Regents*. One such model, proposed by Donna Gitter, compared *Greenberg v. Miami Children's Hospital*, a case in which Canavan Disease families and their disease advocacy organizations protested the commercialization of the research enabled by their tissue and monetary gifts, with the much-examined case in which the disease advocacy organization, PXE International (an advocacy and research organization set up by Patrick and Sharon Terry in 1994 on behalf of their children and other sufferers of Pseudoxanthoma Elasticum—PXE) filed jointly with researchers who isolated the gene for a patent.[73] The sharing in a patent of a disease organization was not the same as individuals having property rights in their tissue after altruistic gifting, but it did at least suggest a departure from the *Moore* majority's rejection of the conversion claim, and its basis in the idea that the biotech industry would be slowed down and public health would suffer if property rights in tissue were granted to owners. Others argued against an ongoing intellectual-property interest in tissue products, but in favor of a moderated, perhaps government-regulated, compensation of (the work of) donation, as something that neither exacerbated the commodification of human bodies nor acquiesced in the inequity of depriving donors alone of all property interests.[74]

BENEFIT SHARING

In California in the first decade or so of the twenty-first century, the new direct-to-consumer personalized genomics promised a new form of autobiography and neo-liberal subject construction; thinking about stem cell research donation from the perspectives of initiatives afoot to bring about open consent or argue for propertization of one's tissue fit in this personalized medicine mold. On the other hand, genealogical, forensic, and conflict zone genomics were invoking collective group identity formation: roots and membership, surveillance and governance, and modalities of reparation and restitution. In short, genomics was emerging as a vital spectrum of technologies for contemporary constructions of and contests over group and personal identity in California and elsewhere. If indeed genomics and regenerative medicine were intertwining, what was true for genomics in the state would also be true for stem cell research. In the Bay Area, where I was working, it was becoming evident that it was going to be necessary not just to include the donor of biomaterial and the taxpaying funder of research in flexible and just ways, but that it was also going to be necessary to think about the bigger social picture of how these biotechnologies were differentially implicating the public. From the forensic CAL-DNA Data Bank in Richmond,[75] to growing support form CIRM in San Francisco for iPSC line banking,[76] to 23andMe in Silicon Valley, the geography of who was included and how in biobanking was rapidly taking shape and it appeared to have in store very different fates for different sectors of our population.

An emerging stratified landscape within California echoed elsewhere in the world, in other regions and nations, and also in the uneven power relations between nations. Internationally, an argument was raging about whether biological material and data should be available as a "commons," as "property," or a combination of, or alternative to, either of these.[77] Studies of demands for benefit sharing and/or group consent for the use of human tissue in scientific research appeared.[78] Visions of meaningful bioequity focused on reform of the appropriation of biological material and information from those with less power to those with more power and suggested reorganization of the give and take of human biologicals and bio-information across borders.[79]

In the course of my fieldwork, several people told me that they thought that it would be right to share profits not only with the funding public via returns to the state for its monetary investment, but also via donations to the non-profit health-care system to increase coverage and reduce disparities in health care. In other words, they recognized two different kinds of benefit sharing that seemed to go hand in hand with public support for Proposition 71. One, returns to the public, was something that CIRM could and did work on with many stakeholders. It articulated policy in accordance with the intent of Proposition 71 in terms of access and affordability to any treatments that were the result of CIRM funding, a state version of the federal "march-in right" the federal government has to grant a new license or revoke an old license on patents arising from federally funded research that are not being developed in the interest of the public, and materials and revenue sharing, including intellectual-property rules that would make sure that both private and public research institutions would pay a share to the state's general fund if they began making a sizeable profit on the products of CIRM-funded research.[80]

The second part of the demand for benefit sharing was not easily addressed, however. To pay adequate attention to human biological justice, as it were, across the state, not only CIRM but also other biobanking and human tissue receiving entities would have to be overseen by a watchdog that could encompass forensic, basic, medical, and biotech research under one roof—a tall order, given the jurisdictional and privacy and proprietary interests at stake. At the end of the beginning of human pluripotent stem cell research in the state, such a public entity had not yet begun to be devised.

IN-KIND RECIPROCITY

A final set of reactions to the need to address the place of donors of tissue in publicly supported pluripotent stem cell research argued for in-kind reciprocity. By in-kind reciprocity, I refer to models that sought to incentivize, empower, recompense, and/or protect donors in ways that were explicitly indexed to the use to which their tissue was being put. My own work fell in this category, and I expressed it at various venues as "allowing donors and their communities to benefit over time from the best the science has to offer."[81] By this, I was particularly interested in tapping into reciprocity flexibly linked to advances in the science itself. Other in-kind models suggested that donors could direct the use of their tissue, could be involved in the governance of their embryo-derived cell lines, or could be charged with helping decide how society should distribute the medical and informational goods from research among all members of the community.[82]

In my policy-relevant work I sought a framework for thinking about reciprocity that would honor the two public gifts to stem cell research—funding and tissue donation—rather than short circuit in over-promising, exploitation, or exacerbated disparities in health and citizenship rights. For this, I worked on a model that would fit the forward-looking state: that is, it would promote research, cures, and the economy through bench-to-bedside, public-private translational innovation (what the majority of the public voted for and donors donated for) while explicitly addressing the crises in the donor-biomaterial-recipient nexus toward which this translational arc could all too easily tend. I sought to incorporate elements, or at least motivations, from each of the approaches to the crisis of reciprocity between donors and recipients discussed above that had been put at stake by Proposition 71.[83] From the comparison with commercial egg donation one could learn about the economic and emotional dimensions of "letting go" of something that could never be entirely let go. From open-consent models came the right to knowledge of one's own data, and the up-front honesty about the ways in which privacy and the informed part of informed consent could no longer be guaranteed. From propertization arguments came a sensitivity to keeping open every part of the value chain, so as not to concentrate money and rights in the hands of the few, and so as not to foreclose short-term or serendipitous economic opportunities. From benefit sharing, one learned to pay attention to group bioethics as well as individual bioethics, and to both infra-national and trans-national bioethics, even while recognizing difficulties in specifying community, benefit, or sharing.

If the donors' investment in the science, as well as researchers' investment, were to be taken seriously, a two-way form of connection would be enhanced. For example, there are substantial benefits to be had both to the quality of the research and to the donor if anonymity is abandoned. Researchers could compare phenotypic characteristics and disease manifestations and trajectories and treatment outcomes with individual samples, while donors could explicitly receive (without directing the research, and not just as an incidental finding) important diagnostic and drug toxicity and other kinds of data from the results of research on their tissue.

I worked on a model that would involve a systematic exchange of information and, where appropriate and possible, a right to the storage of biological tissue and materials such as cell lines derived from those tissues for donors' future therapeutic use as well as researchers' use. The storage itself could be maintained by the government or by other appropriate bodies or even by research institutions themselves, under appropriate storage guidelines. I liked this kind of solution because it matched the nature of the ongoing connection between donors and recipients to the fruits of the research itself. It also offered an incentive to donate that was not purely monetary, thereby making room to critique the commercialization of donation in some cases without sounding the death knell of scientific research.

Stem cell research reciprocity needed to emerge from and contribute to the curation of relations among donors, recipients, data, biomaterial, and intellectual-property rights.[84] The drafters of and voters for California's Proposition 71 intended its potential therapeutic benefits to be available as widely as possible, including to the most vulnerable members of society. Stem cell research in the state needed to begin to spell out over time under which conditions, if any, biological materials and information could be accessed by non-donors, including family members, immigration officers, and the police. And, I believed, the time was right to start tracking the necroscapes of the human biological revolution and paying attention to the ways in which the research might be contributing to stratifying people in relation to the commercialization of human tissue and bio-information.[85] Regenerating bodies and the economy through good science should not be part of a new and unjust triage, but part of a flexible architecture of reciprocity.

THE SUBSTITUTIVE RESEARCH SUBJECT

In this book I have positioned the end of the beginning of human pluripotent stem cell research as a key stage in the advent of personalized medicine. One of the biggest dangers of embracing personalized medicine is to imagine that somehow the social issues of global and local health disparities and excessive biomedicalization go away when medicine and its industries move (in some parts of the world, for some people) to the personalized level. I have tried to show, focusing on various aspects of the ethical choreography of stem cell research in this period, that that is not the case. In particular, I ended chapter 5 with a call to collect and curate tissue and data in a manner that takes reciprocity seriously, and which is equally accessible for forensic and biomedical purposes so as to address the bioscapes and necroscapes explored earlier in the book.

Personalized medicine, targeted to one's own changing genetic, epigenetic, environmental and social circumstances, has an intuitive appeal. It is very exciting (at least to me) to imagine a data-rich future in which curative, predictive, and preventative medicines all take on a more correlated, evidence-based, open-access, minable character. In this chapter, I argue that what I call "substitutive research subjects" are deeply entrenched in the current system of biomedical research and ethics, including in human pluripotent stem cell research, and yet are not well suited to the ethical demands or technical demands of the day. Focusing on animal models, I attempt to show why and how deeply they are entrenched and why they are no longer the best option for the science or the ethics of regenerative and personalized medicine.

I recommend that substitutive research subjects, including non-human animals, be progressively decommissioned from the human biomedical research enterprise as a matter of explicit policy. This recommendation brings with it two immediate questions. First, what could replace animal models and animal testing (and other kinds of substitutive research subjectivity)? Second, what would it take to move biomedical science in that direction—what combination of voices and venues in science could take the lead? I argue that, in a concerted international effort, major science funding agencies, governments, top research universities and the leading science journals, and citizen science groups could together call for and then systematically fund and document and assess an effort to build *in vivo*-ized *in vitro* systems with the express purpose of serving as robust models for research and testing. Over time, the standardization, flexibility, relevance to human medicine, and ethical advantages of such an international effort would lead to its validation and sustainability.

Animals have been used as stand-ins for humans in medical experimentation throughout the history of Western medicine.[1] Vesalius' mid-sixteenth-century work *De humani corporis fabrica* says:

Although I have learned certain things based on the dissection of fetuses and pregnant uteruses and have conducted several demonstrations at the university, I have only used animals, nor have I [studied these organs] with the necessary care, for to this point I have been able to obtain very few women for dissection.[2]

Female animals—or at least their gravid uteruses and their fetuses—were dissected because Vesalius couldn't dissect pregnant women; the animals were an available compromise. The idea that there is a shortage of living and deceased human bodies and body parts for biomedical research, teaching, testing, and treatment has continued to provide a powerful rationale for turning to animals. The second point implicit in the quotation from Vesalius, namely that animals are a scientific compromise, also has a long history. The supply of human cadavers for research has never kept up with demand, even in the present-day United States, where there are functioning for-profit and academic cadaver-procurement services.[3] Surveys of the donation of cadavers and other organs and tissues in recent times suggest the persistence of a variety of reasons why individuals and their families show a willingness or a reluctance to donate bodies, body parts, body systems, or tissue to research.[4]

Over time, the relationship between animals used in medical research and the humans about whom and for whom the majority of the research is undertaken changed.[5] In a standard account, Charles Darwin's work brought a principled basis to the idea that non-human animals and humans were linked in a way that would make the former good biological models for the latter at a deeper ontogenetic level, not just through anatomic similarities.[6] Indeed, as the underlying embryological, cellular, molecular, and genetic processes of life were investigated, animals with few or no anatomical similarities increasingly became seen as good models for humans, and were found to have high levels of conservation across phylogenetic distance.[7] The early twentieth century saw the conscription into science of various model organisms, and their increasing standardization as reliable tools.[8] If the history of using animals for human research is long and deep, so too is the history of efforts to protect—and efforts to protest—the use of animals in research.[9]

In this chapter, I take up the story at the post-World War II point at which non-human animals became an increasingly mandatory *bioethical* substitute for unethical experimentation on human subjects of research. This was part of the same period in which the ethics of experimentation on human subjects became an area of concern and during which a bureaucratic codified basis for the conduct of research began to be established.[10] This rationale for doing research on animals was based on a substitution, and it was as much a departure from as a continuation of the history of doing research on animals in the place of humans. In the period in which I was doing the fieldwork and research for this book, the use of animals in the place of humans for a lot of research and testing was the status quo.

Modern bioethics is often dated from the Nuremberg Doctors' Trial. The trial, *United States of America v. Karl Brandt, et al.*, took place in 1946 and 1947 as part of what is known as the "Subsequent Nuremberg Trials," at which 20 of the 23 defendants were German medical doctors. The defendants were accused of war crimes and crimes against humanity during World War II—of having conspired to carry out, and having participated in, medical experiments on prisoners of war and civilians of occupied countries, and on German nationals, without the subjects' consent. The accusations covered a range of activities, including the mass murder of the Aktion T4 euthanasia program and medical experimentation that involved brutality, cruelty, and torture up to and including death. The ten points of the Nuremberg Code that resulted from the trial included Dr. Leo Alexander's original six points, which he had derived from his observations of Nazi human experimentation during the war and had submitted to the Counsel for War Crimes, and an additional four points that came up during the trial itself.[11] The salience of this code to bioethics has not diminished, even though crucial elements have been added to bioethical statements and regulations since Nuremberg. For example, in the U.S, the demands for distributive justice made in the Belmont Report after the revelations about the persistence of racism evident in the Tuskegee syphilis study and the introduction of institutional review boards are especially noteworthy, as is the major revision of the Helsinki Declaration made in 2000 to include North-South equity in its distributive justice principles. Nonetheless, the original Nuremberg Code still very much undergirds bioethical practice and theory, and Nazi Germany remains the moral nadir and the geopolitical locus of much of the bioethical imaginary.

Of the Nuremberg Code's ten principles, the ones that deal with the requirement that human subjects consent freely, with the necessity that research goals be important and non-random, and with the obligation to minimize harm and to abandon research that has adverse effects on human subjects, are perhaps the most widely known. It is the third principle, however, that codifies the substitutional and sequential logic of animals as research subjects:

3. The experiment should be so designed and based on the results of animal experimentation and a knowledge of the natural history of the disease or other problem under study that the anticipated results will justify the performance of the experiment.[12]

In other words, animal experimentation must be done first to indicate the likelihood that an experiment will work in humans before it is ethical to carry it out on humans. This step replaces humans with animals in the early phases of research, so it substitutes one kind of research subject for another. It also, however, sets up animal experimentation as an epistemological part of the experiment itself, integral to the determination that it is appropriate to proceed to testing in humans. Thus, the logic of animal experimentation in bioethics is sequential, and the epistemology and the ethics reinforce each other. Two things follow from this. First, the substitution of animals for humans makes a fundamental if implicit ethical *distinction* between humans and other animals.[13] Second, the sequential logic makes a fundamental if implicit *connection* between animals and humans: animals must continue to be good (enough) models for humans in whatever system is being investigated, or the sequential epistemological purpose would not be served.

For human pluripotent stem cell research, as for other biomedical research, animals were thought of and used as appropriate biological tools and models for, and necessary ethical replacements of, humans, in the early stages of research.[14] Genetically modified mice—integral to life-sciences research and to biomedical research in general[15]—figured prominently in stem cell research when Proposition 71 passed. Non-human animals were routinely used in stem cell research, including human embryonic stem cell research, from the start. They also featured in several scientific and political debates about the limits of research. This gave animals in the research a second double nature (the first double nature consisting in the sequential, substitutive logic of being both biologically like and ethically unlike humans) of being integral to routine practice yet being invoked when setting boundaries to the research.

Stem cell research, like modern bioethics, had its origins in the post-World War II period, in biophysics experiments on mice designed to assess survivability after radiation exposure.[16] The first mammalian embryonic stem cells were isolated in the mouse. Two research teams were credited with the achievement, one headed by the British scientist Martin John Evans and one headed by Gail Martin at the University of California at San Francisco. Evans' discovery was noted in a 1981 paper, co-authored by Evans and Matthew Kaufman, announcing the isolation and preliminary characterization of mouse "pluripotential" stem cells.[17] Gail Martin, who had earlier worked in Evans' lab, simultaneously but separately published a paper announcing similar results.[18] In her paper, Martin introduced the term "embryonic stem cells" to describe the undifferentiated, self-replicating cells extracted from a very early-stage embryo, and that term gradually entered general usage. Evans also developed the knockout mouse (a mouse in which a particular gene has been made inoperable) to facilitate the study of the genetic control of vertebrate development and disease.

Martin Evans, born in 1942, had done his undergraduate work at Cambridge University and his PhD work in the department of Anatomy and Embryology at University College London.[19] His work on mouse teratocarcinoma self-renewing stem cells was done there, and he was the first to show that these cells could be induced, under the right tissue-culture conditions, to reproduce indefinitely without losing their ability to differentiate. Evans' collaboration with Kaufman at Cambridge University produced the extension of the teratocarcinoma stem cell work to the extraction of embryonic stem cells from normal mouse embryos. This was the work that became the basis of the knockout mouse, which combined embryological and genetic manipulation techniques. Targeted gene knockout has since been used for both gene discovery and for animal models of poorly understood diseases.[20] Mouse targeted gene knockout has become a fundamental tool of molecular and cell biology and the knockout mouse is an embodiment of the substitutional sequential logic of animal models in biomedical research.

The U.S. National Institutes of Health Knockout Mouse Project (KOMP) illustrates the entwining of the mouse model, embryonic stem cell research, and genetics to make and standardize tools for the new biology. KOMP is an "initiative that aims to generate a comprehensive and public resource comprised of mouse embryonic stem (ES) cells containing a null mutation in every gene in the mouse genome." This effort arose out of two international meetings at which "attendees agreed that such a comprehensive resource of null mutants would greatly benefit the biomedical research community and enhance our understanding of human disease." The bioethical and sequential substitutional logic for using animals had become woven into the bench-to-bedside process of translational research. According to KOMP,

mouse genetics in particular exemplifies the translational aspect of model organism research. Through research utilizing the mouse as a model for human disease, investigators can translate basic biological phenomena into a human health perspective. . . . The value of the mouse as a model organism is derived from the fact that the mouse has similar developmental, physiological, biochemical, and behavioral patterns to humans. . . . 99% of mouse genes have homologs in humans. . . . Because . . . of the long history of making and studying mouse mutants, along with the genomic and phenotypic similarities with humans, advance in mouse genetics continue to be a driving force in a broad range of biomedical research activities.[21]

During the first few years after the passage of Proposition 71, pluripotency itself was defined in an animal model. In the George W. Bush era, the U.S. criteria for having successfully demonstrated pluripotency were expressed in terms of the cells' behavior in a mouse blastocyst. The National Institutes for Health maintained, then as now, a registry of all the hESC lines available for federally funded embryonic stem cell research.[22] Before the 2009 change of policy, the NIH embryonic stem cell website contained the following operational definition of pluripotency:

Pluripotency is defined as the ability of a single stem cell to give rise to all of the various cell types that make up the body. Current scientific understanding of pluripotency is based on extensive study of mouse embryonic stem (ES) cells. In mice, the gold standard test for the ability to form derivatives of all three germ layers (ectoderm, mesoderm, and endoderm) is injection of a single putative pluripotent cell (labeled to permit tracking of daughter cells) into a mouse blastocyst. If the labeled injected cell is truly pluripotent, labeled daughter cells will be seen throughout the embryo, in tissues derived from all germ layers.[23]

Early U.S. stem cell lines, however, were also often "contaminated" with non-human animals. Of the 78 human embryonic stem cell lines originally declared eligible for federal funding under President Bush's policy, it turned out that only 22 lines were actually available and of sufficient quality to be used for scientific research.[24] All of these human embryonic stem cell lines were derived using a cellular substrate of mouse feeder cells, however, and were deemed by the Food and Drug Administration to be contaminated and thus not to be usable for treatment or transplant purposes. The question of contamination showed that the interpenetration of non-human animals and human cells was both part of the processes and definitions in question, yet also potentially boundary defining. Contamination raised the risks and ethical problems associated with unintended chimerism. Scientists began deriving human embryonic stem cell lines free of non-human animal contamination.[25]

Once Barack Obama took office as president, in January 2009, prospects for beginning to bring human embryonic stem cell therapies closer to therapies and to market improved dramatically. The first hESC Phase 1 clinical trial to be conducted in the United States was approved by the FDA in late January 2009. As was recounted in chapter 2, Geron subsequently pulled out of the trial. One of Geron's aims had been to effect sufficient standardization and scale-up with human cell lines so that they could eventually be used not only as transplant therapies but also in dish-based drug discovery and toxicity testing. This may have had the potential to reduce the numbers of animal or human subjects in the future. Despite the fact that this was, arguably, a step in a gradual paradigm shift away from animal and human testing regimes of the past, the application itself only served to illustrate the extent of the involvement of animals in the extant regime. For the clinical trial in question, Geron submitted to the FDA a 21,000-page IND (investigative new drug) application that "described 24 separate animal studies," including "studies that showed no evidence of teratoma formation 12 months after injection" "into the injured spinal cords of rats and mice," and studies that "documented absence of significant migration of the injected cells outside the spinal cord" and absence of "systemic toxicity or increased mortality in animals," as well as "significantly improved locomotor activity and kinematic scores of animals with spinal cord injuries" and "improved axon survival" upon histological examination.[26]

The first proof of principle that iPS cells could lead to live births and so were functionally equivalent to ES cells was carried out on mice. IPS cell technology, if successful in humans, held the potential to make it practicable to grow a patient's own tissue in the lab for personalized testing and for the patient's own regeneration, and to make it practicable to avoid sacrificing a human embryo. The mouse studies in question used as their starter somatic cells mouse embryonic fibroblasts, which were exposed to the four "Yamanaka factors" (transcription factors that induced reprogramming). The resulting cell lines had their pluripotency tested in tetraploid mouse blastocysts that would not have had the ability to grow a fetus without the pluripotent cells. One of the authors of one of the early iPS studies, Zeng Fanyi of Shanghai Jiao Tong University, echoed the NIH language of the "gold standard" proof of pluripotency in a press briefing in July 2009 when referring to the birth of live mice from iPS cells.[27]

In short, at the end of the beginning of human embryonic stem cell research, the ethics and best scientific practice that each dictated doing things in animals first were intimately entwined. Even if everyone agreed on a goal of testing less on animals, and of stem cell research and personalized medicine and dish-based testing as inherently part of that goal, it would be neither easy nor a simple story of one way of doing things being good while the other was bad. The ethical and scientific genealogies of existing experimental practices would have to be reverse engineered to an earlier stage and rerun, or grafted onto new ways of doing things.

Aside from all the ways in which animals were ethical substitutes for and scientific models and tools of humans in hESC, animals were also boundary defining in U.S political culture. The boundary between human and non-human has long been contested, protected, and crossed in bio-medicine and its political and cultural representations.[28] Non-human animals played an especially important boundary-defining role in a U.S. consensus to say no to human reproductive cloning and to germ-line human-animal chimeras.

Somatic cell nuclear transfer cloning involves fusing an enucleated egg cell with the nucleus of an adult differentiated cell and inducing the result-ing embryo-like entity to divide and differentiate. The resulting offspring, if brought to term by being implanted in a uterus, is a nearly identical genetic match not to the egg donor or the animal carrying the pregnancy but to the donor of the adult nucleus. Thus the offspring has only one major genetic "parent" despite dividing the task of biological reproduction into three (the egg donor, who contributes the mitochondrial DNA but no nuclear DNA, the nucleus donor, and the gestational surrogate), and is thus a kind of clone of the nucleus donor. This SCNT procedure is known as "reproductive cloning." The birth of the first mammal conceived by somatic cell nuclear transfer cloning was announced in 1997. Dolly the Sheep became one of the world's best-known animals, and the lead scientist on the effort, Dr. Ian Wilmut, also became famous overnight.

For Dolly, the adult cell came from a sheep mammary cell. (The name Dolly made reference to the singer Dolly Parton, known for her ample bosom.) Dolly's birth was a proof of the principle that the nuclei of already differentiated somatic cells could revert to embryonic status and give rise to all the tissues of the body; it also showed that reproductive cloning was a reality.[29] Wilmut, a pioneer in gamete and embryo freezing and cloning, is credited with a number of other firsts in the field, including the birth in 1973 of Frosty, the first calf born from a once-frozen embryo. Wilmut's team soon realized that the SCNT procedure could be used to derive embryonic stem cell lines. The promise of histocompatible embryonic stem cells became a major motivation for extending this research, and Ian Wilmut's team (at the Roslin Institute and the University of Edinburgh) was granted one of the licenses given out by the U.K's Human Fertilisation and Embryology Authority to carry out human cloning research. Wilmut also became scientific advisor to Geron Bio-Med, owned by Geron Corporation, which operated the licenses from the SCNT patents pertaining to Dolly.

Public stereotypes and fears of cloning took attention away from other ethical and safety aspects of nuclear transfer and its role in regenerative medicine. One aspect was the biological profligacy: Dolly was the only successful birth from an original 430 enucleated eggs. In view of the risks of procuring human eggs, this low yield, which appeared to be typical of attempts to clone mammals, was widely considered to be unacceptable for human research. Likewise, Dolly was put to sleep at the age of 6 years after showing signs of premature aging; the biological uncertainties surrounding tissue made by somatic cell nuclear transfer cloning for many underscored the unacceptability of human reproductive cloning. In addition to these technical concerns, popular representations of human cloning abounded. Despite claims to the contrary (some of them made by the pro-cloning Raelian cult), the reproductive cloning that resulted in Dolly's birth had never been shown to work in humans and may never even have been tried. In many countries human reproductive cloning was illegal— indeed, it is striking how strong the consensus against human reproductive cloning is—even as it became more routine in the reproduction of agricultural animals, police dogs, and pets, and subject to controversy in organizations such as the American Kennel Club.

Rather than reproductive cloning, scientists wanted to extract patient-specific embryonic stem cells using this procedure. This held out the promise of creating human biological therapies from histocompatible cell lines for individual patients, and of mitigating graft-versus-host disease that plagues transplant medicine. Because there would be no reproductive intent or outcome at any point, this procedure was called "research cloning" or "therapeutic cloning" by some, to distinguish it from reproductive cloning. This procedure—deriving embryonic stem cell lines from embryo-like entities after SCNT—hadn't yet been shown to work for human ES cells. As was noted in chapter 4, the highest-profile scandal of the hESC era occurred when the Korean stem cell scientist Hwang Woo Suk and his team at Seoul National University were forced to retract papers in which they claimed to have created just such patient-specific embryonic stem cell lines.[30]

The passage of Proposition 71 permitted research cloning (SCNT hESC) while explicitly banning reproductive cloning. Proponents of the research tended to downplay the word "cloning." The combination of the liminal or boundary-defining nature of reproductive cloning (acceptable for animals, not acceptable for humans) and its proximity to research cloning, the Hwang scandal over SCNT, the rise of iPS cells and the potential that it could render SCNT redundant, and the problems with the recruitment of egg donors in a non-reproductive context for such research have all undermined research cloning. If SCNT were to be shown to be successful in humans this might change, but until then it is an example of how stem cell research paces out the boundary between the human and the animal, just as importantly as tying the two ever closer together.

It is important to note, however, that human reproductive cloning, which seems so clearly and by such strong consensus to provide a limiting case of what is (often) acceptable in animals but not in humans, is more complicated on further examination. Consider how Ian Wilmut and his colleague Keith Campbell negotiated the question of human reproductive cloning in a book documenting their feat in creating Dolly. In a display of perhaps peculiarly British scientific understatement, Dolly's lab fathers present themselves as run-of-the-mill agricultural scientists who just happened to have been interested in the question of the potentiality of adult mammalian epithelial cells, and thus to have been thrust into the maelstrom of public outcry and politics. Their amanuensis for the book, the natural history and biodiversity conservation writer Colin Tudge, solicited their views on cloning. The result was a book titled *The Second Creation: Dolly and the Age of Biological Control*.[31] Despite the far-from-modest title, the book was staunchly in the genre of science writing in the service of the public understanding of science. It displayed several characteristics of that genre. They appear to be saying the following:

We're just scientists; let us tell you why the science in question is exciting. The media and popular culture distort what we do, so let us pinpoint the misapprehensions that have been promulgated.

Although we cannot understand why anyone is asking us about things like cloning people, everyone should just say no to applications of the technologies that don't answer sensible scientific questions or solve dire medical problems.

When we seem to engage in less than purely scientific endeavors such as corporate partnerships and patent activity, it is all actually pretty routine in science these days, given tight budgets and the grant dependence of universities, so don't worry about our impartiality.

Despite the formalities of the genre, however, the treatment of human reproductive cloning was highly instructive. If one took Wilmut and Campbell at their word, although they knew more than anyone else about the science of cloning, they had no particular claim to expertise in the social and ethical aspects of mammalian cloning. One's mother or priest or friend or physician would give counsel that was at least as good. This may be so in some important sense, but in another sense they were in a unique position to comment on social meanings of cloning: people contacted them. After their technical abilities were made public in the physical being of Dolly, members of the public and specialists alike sought the Roslin team's advice as to the technical feasibility and moral judiciousness of a myriad of imagined scenarios. Some people were moved to contact Roslin to ask about the availability of animal *and* human cloning services, thereby showing that in some circumstances some people are all too able to make the jump from animal to human reproductive cloning, and for the most "human" of reasons: longing, bereavement, despair, desire. For example, in a chapter attributed to Wilmut one finds the following discussion of requests to clone deceased loved ones:

I have often been asked—although I dread the question—whether it is possible to re-create some dead loved one. I was first asked this over the telephone from another country. Over the telephone or in the lecture hall, it is not possible to say much more than "No, at this time it is not possible." . . . The more expanded answer, however, is that there are two caveats. . . . A child (or cat or dog) that was cloned . . . would not simply be a facsimile of the original. . . . And . . . how would the parents avoid making comparisons with the dead brother or sister?[32]

Wilmut only hints at what the emotional content of a telephone call must be like in which he is asked if he can replace a son or daughter killed prematurely from illness or in a freak accident. He dreads the question (he refers to this dread a couple of times in the book). How can he turn them away with a scientist's detachment, but what solace can he responsibly give? He also presumably realizes that the stock technical answer that "humans cannot yet be cloned" is simply not an answer to the longing expressed in the imaginative leap that took the parents from Dolly the sheep to a living, breathing replacement for their lost child. It must have been at least somewhat tempting to answer in another register altogether, to meet desperation not just with compassion but with scientific passion, and to commit to the deployment of the technology for humans, or at the very least to gesture to a "holding pattern," such as the storage of DNA from the deceased child. After all, this translation of compassion to scientific passion is commonly rehearsed and often highly effective in science-patient advocacy collaborations. The reluctance to do so indicates that, for now, the line is still holding on reproductive cloning.

Both Ian Wilmut and Keith Campbell have also been asked about cloning by individuals and couples struggling to overcome infertility. Similar to the desire to replace dead loved ones, infertility is often experienced as an excruciatingly embodied, moral, and personal crisis. Cloning is considered in cases

when one person alone seeks to reproduce, without a sexual partner; or when two partners either fail to produce gametes at all or produce incompatible gametes—as is the case with homosexual partners. . . . Male homosexual couples might conceivably be cloned with the aid of egg donors and surrogate mothers, while female couples could be far more independent. . . . One member of a lesbian couple might provide a nucleus from a body cell, and the other could provide the cytoplasm; next time around they could reverse the procedure. Of course, by such means, a lesbian couple could produce only daughters. A woman could clone herself precisely if one of her own nuclei was introduced into one of her own enucleated oocytes. Many combinations can be imagined.[33]

Just as they did for the encounter about cloning deceased loved ones, Wilmut and Campbell tame the unruly reproduction using familiar arguments. "It is certainly not up to us to comment on where the limits of parenthood ought to lie," they say. Instead, they refer such involuntarily childless individuals and couples to already existing, and thus presumably less problematic, reproductive technologies, such as gamete donation and surrogacy. They also remind the reader that adoption is available when Campbell laments the genetic essentialism inherent in the desire to have genetically related offspring by asking "Why is genetic relationship considered to be such a big deal?" Talking about possible scenarios in which cloning might be almost reasonable tames it somewhat. The landscape of the intimate is powerful at rendering things familiar, making the extraordinary curiously ordinary.

Repairing loved ones through regenerative medicine got a different treatment from Wilmut and Campbell. As was common within the ambit of stem cell research, the authors separated out germ-line gene therapy from all other genetic engineering as the only kind that "raise(s) outstanding ethical issues." Wilmut and Campbell flagged the dangers of manipulating the germ line in an "experimental" fashion, citing the threat to the well-being not just of the individual in question but of all future generations. Single-gene defects and somatic gene therapy, however, were treated as falling within the near medical horizon, and were assimilated to other imminent breakthroughs in medicine: problems remained to be solved, but they were technical ones. "We can," Wilmut and Campbell wrote, "draw the traditional distinction that runs through all medicine: on the one hand, the correction of obvious disorder; on the other, the enhancement of what already works reasonably well. . . . The distinction is not always easy in practice, but in principle it seems clear enough." This view resonated widely. For example, in 2000—the year Wilmut and Campbell's book came out—Christopher Reeve argued before the Senate's Subcommittee on Labor, Health, and Human Services Committee on Appropriations that "no obstacle should stand in the way of responsible investigation of their possibilities . . . to cure diseases and conditions ranging from Parkinson's and multiple sclerosis to diabetes and heart disease, Alzheimer's, Lou Gehrig's disease, and even spinal-cord injuries like my own."[34] In short, it was possible to argue for fundamental boundaries around what is acceptable for humans in principal (cures, but not germ-line genetic modification or cloning), even if the boundary was in practice hard to pin down, and even while illustrating through vignettes the potential of those boundaries to be reconstructed or breached when the technologies hold out hope for actual people living real lives.

Cloning did not form the only animal-human boundary zone. While the United States was grappling with the Bush policy and the implications of abortion politics for stem cell research, stem cell research in the United Kingdom was developing in a more permissive regulatory environment. When I visited stem cell scientists in the U.K. in 2006 and 2008, I went to meetings and spoke with some of those involved in promoting the cytoplasmic hybrid embryo ("cybrid") idea for deriving embryonic stem cell lines in a way that might get around some of the field's procurement obstacles. Working in an environment where SCNT was legal with a license, the idea was that scarce women's eggs needed for somatic cell nuclear transfer could be replaced by eggs from non-human animals obtained from abattoirs. These eggs (cow eggs were proposed in the U.K. case) would then be enucleated, and the nucleus of a human somatic cell would be introduced and induced to start dividing. This would enable researchers to derive patient-specific human embryonic stem cell lines without needing a woman's egg to do so. Two people (who didn't identify themselves or their research to me as feminist) suggested to me that it was "feminist," in that it would relieve pressure on women to donate eggs. Although the U.K. had a long history of conflict between science and animal rights, its regulator of human embryonic stem cell research, the Human Fertilization and Embryology Authority, had, after a lengthy public comment process, allowed researchers to apply for licenses to carry out cybrid research.[35] Whatever the reality of public opinion, and despite the HFEA's having issued three cybrid licenses, cybrid work had fizzled out by the end of 2011.[36] Stephen Minger, the U.S.-educated face of British cybrid research advocacy and the leader of one of the two groups that got the first HFEA licenses for it, left Kings College London for the private sector. In an interview published in *Nature* in August 2009, he cited as part of his motivation the desire to move drug toxicity testing out of animal models and into human cells, for scientific, clinical, consistency, and scale-up reasons:

Most of the screening is done using animal cells . . . or is done using human tumour-cell lines that don't faithfully represent true primary human cells. . . . The power of using embryonic-derived cells is consistency, both in terms of quality and genetic background.[37]

Meanwhile, in the United States, the possibility of the introduction of animal cells into humans or human cells into non-human animals in ways that affect the germ line and can be passed on to future generations continued to fuel fears about stem cell research and the anti-abortion lobby's opposition to the research. This concern for the integrity of the human species made its way into the voluntary guidelines for human embryonic stem cell research that the National Academy of Sciences issued in 2005.[38] It showed up again, unaltered, in the 2009 NIH Guidelines. The Obama NIH guidelines dealt in every other particular with the provenance and procurement of eggs, embryos, cells, and "acceptably derived lines" for stem cell research that could be federally funded. There were only two exceptions, where the guidelines considered the *disposition* of embryonic stem cells. They both dealt with animal-human hybrids or chimeras:

III: Research Using Human Embryonic Stem Cells and/or Human Induced Pluripotent Stem Cells That, Although the Cells May Come From Allowable Sources, Is Nevertheless Ineligible for NIH Funding.
A. Research in which human embryonic stem cells (even if derived according to these Guidelines) or human induced pluripotent stem cells are introduced into non-human primate blastocysts.
B. Research involving the breeding of animals where the introduction of human embryonic stem cells (even if derived according to these Guidelines) or human induced pluripotent stem cells may have contributed to the germ line.

At the end of the week in which the NIH Guidelines were finalized, Senator Samuel Brownback (R-Kansas) introduced the Human-Animal Hybrid Prohibition Act of 2009.[39] Senator Brownback (who had introduced the 2004 pro-life Unborn Child Pain Awareness Act and the 2005 Human-Animal Chimera Prohibition Act) was joined by Senator Mary Landrieu (D-Louisiana) in introducing the bill, with co-sponsorship from nineteen other Republican senators, including John McCain of Arizona, who had run for president the year before. In introducing the legislation in July 2009, Senator Brownback called it "both philosophical and practical as it has a direct bearing upon the very essence of what it means to be human, and it draws a bright line with respect to how far we can go." Yet, bright line or no, the legislation in fact made a case to allow all sorts of human-animal mixing to go on, including genetic engineering, but drew the line at "human-animal hybrids," "defined as those part-human, part animal creatures, which are created in laboratories, and blur the line between species" by making "changes in the germ-line" that can be "passed along through the gene pool to the rest of humanity." This kind of germ-line genetic modification was contrasted with "legitimate research" in health care and with "genetically engineering safe products and herds" for humans in agriculture.[40] In the United States, the positioning of these lines varied by political persuasion. This Republican bill asserted that the line between human and animals is essential to human dignity, while seeking to continue support for genetic engineering in agriculture, and the use of animals in health-care research. The section below on animal politics gives an example of a Democratic nominee who drew these lines in somewhat different ways. Nonetheless, the phenomenon was the same: animal–human hybrids were both the feedstock of research in and for humans, and set the ground for (constantly shifting and contested) boundary maintenance as to the difference between humans and animals, and what was thus to be considered ethically acceptable.

HUMANIZING THE ANIMAL MODEL

In order to be useful for biomedical research for humans, an animal model has to be sufficiently human in relevant ways. What that means depends on which animal model is used and in what the research consists. Over time, animal models have been refined to work better as tools—that is, to be more consistent or standardized. For example, the knockout mouse project described above is a public repository and active collection effort for an animal-model series of embryonic stem cell lines derived from mice embryos each of which is missing a single gene at a different defined locus ("null alleles"). The project has a preference for a particular inbred strain of mouse that has been developed for this knockout work: C57Bl/6, or "black 6." For purposes of research, the relevant human gene can be substituted for the missing gene. This is one example of "humanizing."

Humanization happens in a variety of ways, not all of them transgenic in nature. Consider the SCID-hu mouse, a severely immunodeficient mouse that has had parts of the human immune system transplanted into it and is used as a preclinical model system for diseases where the properties of the human immune system are essential if results are to be meaningful. For example, SCID-hu mice are used in work on the human immunodeficiency virus (HIV), which doesn't grow well in mice.[41] SCID mice have also been transplanted with human hematopoietic stem cells.

At first sight the concept of humanized animal models seems illogical and unethical: have an animal stand in for a human; if it can't, then alter it until it can, even if that means making it part human and/or making it suffer from a deadly disease to which it was previously immune. In the absence of alternatives that can give the kind of information of a living system ("*in vivo veritas*," as the saying goes), however, and in view of the human suffering and death toll of various diseases, it is little wonder that many sufferers and many people involved in research, patient care, and cure approve wholeheartedly of animal research as long as it was is within stringent standards of the "three Rs" (refinement of treatment, replacement with in silica or in vitro or other alternatives where possible, and reduction of numbers of animals used) of animal use and care.[42] At a certain point, however, the level of humanization needed becomes so extreme that it seems that the effort could be directed instead to animating *in vitro* systems. This is especially true of stem cell transplantation, which requires extensive alterations (for example, introducing human cytokine genes into highly immunodeficient mice) to humanize mice enough so that they can tolerate human xenografts and support human cell and tissue growth and differentiation in scientifically and clinically useful ways.[43] Animal models were already known to have significant flaws in areas in which stem cell research had great promise, such as drug toxicology testing (for example, human liver and heart toxicity of drugs), and to be slow and expensive.[44] At what point did animal models become bad science, and stem cells become part of the way out rather than a field of research needing ever-more-heroic humanization?[45]

As a member of two Stem Cell Research Oversight Committees (SCROs), I watched the involvement of animals in human stem cell research evolve over the period covered in this book. In California, after their establishment in the wake of the passage of Proposition 71, SCROs moved fairly quickly to minimize meeting times and the regulatory burden of their activities on stem cell researchers. One means of doing this was to make sure that elements of research oversight that could be reviewed by one or another of the pre-existing review committees—IRBs (institutional review boards), responsible for protecting human subjects; IACUCs (institutional animal care and use committees), responsible for animal care and stipulating the conditions for their use in research; and biosafety committees—were handled in those committees. This implicitly divided up topics of ethical concern that are involved in stem cell research into the generic ones (shared with other life sciences) of the protection of human and animal research subjects, and biosafety, leaving SCROs with embryos, gametes, and hybrids to take care of. In a sense, this reverted back to the original debate about embryonic stem cell research in the United States, according to which embryos might have to be treated as their own kind of research subject needing a committee as much as animals and humans. At first, while the challenge from pro-life groups was still a threat to releasing research funding, SCROs encouraged more open ethical inquiry.

Quickly, however, the meetings closed in on regulatory compliance (and relief from regulatory burden). By the time Obama effectively lifted the specter of having to consider the possible research subjecthood of embryos, the question of whether there was a need for SCROs at all had begun to be raised. It was, for example, explicitly discussed at a California Institute for Regenerative Medicine meeting on SCROs held at the end of June 2009 at which I was a participant.

Where animal testing and animal models were concerned, this normalization meant separating research uses of animals into those aspects of routine research (such as gold-standard pluripotency assays) that required IACUC insight or oversight but didn't require SCRO insight or oversight and strictly regulated chimeric situations that required the highest level of SCRO oversight:

These amendments state that CIRM-funded purely in vitro research utilizing covered stem cell lines or the reprogramming human somatic cells with the aim to derive or create a covered stem cell line may not commence without written notification to the designated SCRO committee [the lower oversight option]. Research may include animal assays to evaluate pluripotency; however, subsequent introduction of derived covered stem cell lines in non-human animals shall be reviewed in accordance with section (e) [full committee oversight]."[46]

In sum, part of the rationale for regulatory burden removal invoked the need to "facilitate coordination among the SCRO, IRB, and IACUC."[47] Restricting SCRO review to work involving human embryos made sense if SCROs were functionally equivalent to IRBs and IACUCs, each accounting for its own kind of living research subject and each with its moral injunctions ("autonomy, beneficence, and justice" for post-natal human subjects, "care and use" for non-human vertebrates, "consented and acceptably derived" for embryos).[48] Each would cover a domain of vulnerable life that (at least according to some) was due special consideration, and each was potentially vulnerable to misuse/mistreatment in research. Thus, whereas IRBs protected human subjects' rights, and IACUCs looked after animal welfare, SCROs, seen in this light, were responsible for human embryos and for the prevention of unacceptable reproductions such as chimeras. Each of these committees had its own history.[49] The trouble with this organization was that it left ethical concern about animals outside stem cell research oversight. This, in turn, left no opportunity to think through the very idea of the involvement of other lives in this research—something that wasn't going to be done in stem cell research oversight committees.

ANIMAL POLITICS

POSITIONS ON THE STATUS OF ANIMALS

What would it be to move to beyond the discrete committees of research with living beings, and to reconsider entwined research subjecthood in stem cell research? Two weeks before the 2009 NIH Stem Cell Research Guidelines took effect, President Obama's pick to head the Office of Information and Regulatory Affairs (under the Office of Management and Budget, in the Executive Office of the President), the Harvard University law professor Cass Sunstein, had his appointment temporarily blocked by Senator Saxby Chambliss (R-Georgia). Chambliss, the ranking member of the U.S. Senate's Committee on Agriculture, Nutrition, and Forestry, warned that Sunstein believed that "animals ought to have the rights to bring lawsuits" and warned of "threats of frivolous lawsuits" against "hard-working farmers and ranchers in America."[50] Senator Chambliss removed the hold on July 16, 2009, having been persuaded that Sunstein would protect agriculture and hunting. In 2004, Sunstein had made the case that animals should be permitted to bring suit (represented by humans) when faced with violations of animal-protection law.[51] This was part of long tradition of arguing for legal personhood in the sense of "standing to sue" for entities other than individual humans, according to which companies already have standing to sue in the U.S. legal system.[52] This position did not say anything about new animal-rights legislation, but instead would allow private individuals to bring suit on behalf of animals to aid enforcement efforts already in place. Nonetheless, the position was widely interpreted as a partisan attack on farmers, ranchers, hunters, and the Second Amendment.

At the same time, efforts were gathering steam to tighten up the requirements on pet ownership in several parts of the United States. The city of Los Angeles, for example, had adopted a new law in 2008 that required all cats and dogs over four months old to be spayed or neutered; exemptions for show dogs, service dogs, working dogs, and breeding dogs and cats were written into the law, requiring registration, various permits, and insertion of microchips. A similar state-wide bill, Senate Bill 250 (Dogs and Cats: Spaying and Neutering), would have come up had it not been for the financial crisis. If one believed the text of Los Angeles Ordinance No. 179615, and the minutes of the Spay/Neuter Advisory Committee that was appointed under the Ordinance, the chief concerns of the LA Animal Services and the Los Angeles law and other similar ones were how to reduce the ever-rising euthanasia and animal abuse and abandonment rates, how to keep the escalating municipal costs for care of abandoned animals down amid financial meltdown, how to make preventative spaying and neutering procedures accessible and affordable to all pet owners by offering low-cost procedures and vouchers, and how to ensure compliance.[53] Among those most strongly opposing the bill were a subset of those who kept pets—hobby breeders, for example—and spokespeople for the meat industry and ranchers.

To many owners of animals in the exemption categories, such as hobby breeders of pure-bred dogs, the Ordinance and other similar efforts nationwide were thinly veiled attempts to erode the right to (or the concept of) animal ownership by making it prohibitively difficult, expensive, and time consuming to get all the necessary paperwork; furthermore, those proposing such legislation were seen as having the extremist cast of animal-rights activists, advocating a kind of animal rights that is perceived as negating the rights of owners to live in particular ways with their animals. The result is, to the lay reader, a surprising polarization of views between two groups of animal lovers:

The unfolding situation in Los Angeles gives dramatic and clear proof that the ultimate goal of all animal rights-inspired legislation is a step toward the elimination of animal ownership in America. . . . For dog owners, it means that any compromise with animal rights activists is illogical, unwise, and totally illusory. . . . The only alternative is to fight back courageously against all animal rights legislation, and refuse to quit, surrender or compromise.[54]

How can one make sense of this polarization between animal-rights advocates and forms of livelihood and leisure involving animals? What is the underlying landscape of animal politics, how culturally specific is it, and how does it affect animals as research subjects?[55] As I did above for the bioethics of research with animals, I shall take both a snapshot approach and a genealogical approach to sketching politically salient positions on animals in California during the period covered in this book. A rubric that divides positions into "animal rights" perspectives and "animal use" perspectives works fairly well in the U.S. political context. I shall also briefly discuss "animal conservation," "animal welfare," "animal mind," "animal

identification," "animal naturalization," "animal salvation," and "animal den-
igration," all of which are also salient to the animal research politics I turn
to in the next section, and which imperfectly fit the rubric of use versus
rights.

"Animal use" positions encompass a wide array of ways of living with
animals where animals form the basis of one's livelihood and/or one's
capital expenditure. This ranges all the way from circus animals and such
things as dog fighting, horse racing, and hunting to working animals like
sheep dogs and the proverbial canaries in coal mines and lab mice, to agents
of bioterror and biosecurity, to the consumptive side of pet ownership, to
the legions of animals kept for their meat, leather, or other products.[56]
Meat production itself ranges from backyard chickens to huge meat-production facilities.[57] These different "animal use" worlds overlap but can also
be very distinct from one another, not only in terms of how mainstream
they are with respect to the law but also in terms of the gender, race, class,
and affective content of the human worlds in which each figures. The
world of elite horse racing and that of dog fighting, for example, despite
both being competitive higher vertebrate activities capitalized through
gambling, are quite different.

"Animal use" positions are often based on the dominion of humans over
other animals, but also run the gamut to versions of a co-dependence ethic,
where the livelihoods of both the humans and the animals in question are
seen as intertwined. For example, whereas some people have no qualms
about eating commercially produced meat without considering the conditions of meat production, others prefer to eat only meat from animals that
have been raised in humane ways and/or in ways that do not compromise
species survival or environmental goals that allow non-human animals to
survive. There can thus be a big philosophical distance between these two
ends of a spectrum on use. "Animal use" positions also tend to emphasize
the historical, nationalist, and economic (and often masculinist and rural,
and sometimes also colonial and racist) relations between animals and their
humans. Some "animal use" positions are strongly associated with Republicans in the United States, especially through hunting and ranching, and
with certain conservative Christians who believe that dominion over
animals is a matter of faith. There are profoundly secular "animal use"
strands, however, evident in such things as markets for fur and leather goods
and other value-added animal products whose rarity translates into status.

"Animal rights" perspectives, on the other hand, start with the position that animals should be valued for their own sake and not for their usefulness or value to humans. Many hold that the best way to do that is to accord them rights-bearing status so that they are protected under the law and in international treaties. Unlike "animal use" perspectives, "animal rights" perspectives tend to start with the philosophical and derive the practical, rather than the other way around. Exactly what rights different proponents think animals have, the basis of those rights, and the means of promoting and protecting those rights varies. Some believe that non-human animals should be as protected under the law in the same manner as humans; others even believe that some animals have more rights than some humans; yet others do not put animals and humans on the same scale. Practically, "animal rights" positions range from the positions of those who would emancipate all animals currently being used in any kind of service to humanity to the positions of those who want to maximize the humane treatment of animals and avoid vivisection and unnecessary killing. For example, some vegans and other proponents of animal rights oppose eating or wearing any animal products or using or profiting from animals in any way, including pet ownership, whereas many vegetarians merely oppose unnecessarily sacrificing animals for food. Again, there can be a big distance between these different rights perspectives. Although the emphasis is on liberating animals from being used unnecessarily for economic and other human ends, there are still significant economic and emotional stakes in "animal rights" perspectives. Opposition to the use of animals in research commonly gets portrayed as a view held by the violent liberationist fringe of the animal-rights community, but that is probably because it is violent threats or attacks on researchers or research facilities that involve the police and people's safety that get reported. There is a more mainstream variant of the "animal rights" position that sees the successive enfranchisement over time of more and more contenders to subjecthood, and imagines non-human animals as the next class of beneficiaries of this Enlightenment trajectory. I address some of the strengths and some of the weaknesses of this view below.

Animal conservation has many use variants and many rights variants, but generally it argues for the intrinsic ecological worth of individual species and offers a plethora of reasons why conservation in general or the preservation of particular species is good for the economy, for society, for the environment, for morality, and so on. From zoos (which themselves range from imperial menageries to genetic "Noah's arks") to *in situ* wildlife conservation to non-consumptive wildlife tourism to the Endangered Species Act to international trade in endangered species, animal conservation covers an enormous range of rationales and values for non-human animals.[58] Its ethical positions include North-South equity, the need to help poorer countries support fauna that is already extinct in rich industrialized nations, promotion of tourism, conservation of biodiversity, various environmental ethics that de-center "man" and reconnect humans to nature, and ecofeminist perspectives that equate the domination of nature by humans with the domination of women by men.

"Animal welfare" unlike "animal rights," is an expression that is often used to describe those whose jobs involve caring for and improving the lives of animals, even or especially when they are serving or working with humans. It is an ethic of care and concern promulgated by those who live with animals that is pitted against the perceived abandonment of animals by proponents of animal rights. For some proponents of animal welfare, advocates of animal rights make a fundamental mistake of species-centrism if they fail to see that what looks like freedom to them could be a death sentence to many animals. As the "animal welfare" position is articulated among such groups as hobby breeders, for example, it provides a defense of pure-bred animals but also a basis for condemning puppy mills and other human manipulations of animals that put economic gain above the love and care and skilled breeding of the individual animals. The autistic scholar Temple Grandin advocates the introduction of the humane treatment of animals even into the largest meat-production facilities, believing that even if the animals must die it is still important and possible to be empathic and minimize their pain and suffering. This ethic of the carer—the farmer or rancher who knows and cares best for the animals that are his or her livelihood, or the pet owner who would do anything for her or his pet—can be compelling, even though it doesn't answer the deontological demand of some animal-rights advocates never to use an animal as a means to another end.

Another important set of currents swirl around the question of how animals' minds differ from human minds. For some, the boundary between human and non-human animals is empirical rather than *a priori*, and so it is amenable to experiment. Over time, criteria that supposedly separate humans from all other species—tool use, self-recognition, transmission of culture, language use, interspecies relations—have fallen by the wayside upon investigation. Questions about how clever or self-aware animals are have an impact on how much they can suffer and what kinds of dignity and freedom they should be afforded. It is probably true that we have historically erred on the side of attributing too little sentience to non-human animals; it is also the case that for some people it makes no difference where the line lies; it is the fact of the difference, whatever it might consist in, that underlies their humanism. For still others, the point is to treat animals like humans through an identificatory or empathic attachment to them; for others, animals play a cleansing or salvational role, being seen as morally innocent and thus superior to postlapsarian cruel, degraded, and warmongering humanity. Sometimes animals serve to naturalize, or make normal and acceptable, what had previously been seen as deviant. For example, showing that homosexuality is common in the animal world has been a common way to naturalize and thereby destigmatize what in some quarters has been seen as deviant behavior: if animals can be homosexual then it must be natural, and, there is probably nothing that can be done about it or no need to do anything about it.[59] As a caution against the romance of the destigmatizing power of naturalization, critical race scholarship has suggested that the search for animal mind echoes the violent history of attributing lower mental capacity and a lack of rationality to slaves and others.[60]

Positions akin to the opposites of the last two positions often surface, however, and these are positions of the most importance to bioethics. On the one hand, holding non-human animals up as prelapsarian can lead to a preference for (some) animals over entire classes of humans by some powerful groups or spokespeople. Adolf Hitler's affection for his German shepherd Blondi and the centrality of animal protection to Nazi propaganda is the limiting case. The prominent utilitarian ethicist Peter Singer has argued strongly for animal rights, but also has implied that some animals may be more worthy of life and resources than severely ill or disabled humans.[61] In the case of Hitler, it was not the love lavished on his dog but its banal and chilling aspect when considered alongside mass genocide that perhaps most shocks. In the case of Singer's utilitarianism, it is also not the value granted to animals but the arrogance and violence of assuming anyone has access to the metric according to which different human lives' worths are measured that causes the most moral disagreement, as groups such as Not Dead Yet have pointed out.[62]

Degrading or denigrating some groups of humans by labeling or treating them (or us) as if they (or we) were animals is also relevant to bioethics. Though it is sometimes destigmatizing, or morally elevating through the kind of naturalization or prelapsarian cleansing mentioned above, to say that something is animal-like, there is an equally important opposite use. Thus, to "treat someone like an animal" means to treat her or him precisely as if lacking in (or in the act depriving her or him of) whatever is constitutive of humanity. This trope of animality is at the heart of biopolitics: under what conditions do some people treat other people like animals, thereby cementing the meanings of power (those able to reduce the other to the animal), abjection (the state of being so reduced), and the modern state (the political entity wherein the legitimate use of power encompasses actions that sort the powerful from the abject)? Modern bioethics arose out of Nazi biopolitical abjection, and the morally and epistemologically embedded model animal was the means whereby the abjection of those experimented upon *as if they were animals* was diverted onto the bodies of literal animals. Legitimate state power became prosecuted war crime at the moment at which animals became the legitimate subjects to be treated like animals by science.

If the concentration camp provided the original site of biopolitical abjection for contemporary bioethics, over time other genealogies of biopolitical abjection have yielded their own principles that have been grafted onto the Nuremberg principles to form our current bioethical instrumentation. The plantation and slavery also "treated humans like animals," using a somewhat different racial logic that employed the politically diffuse and civilian practices of (pseudo)science and commerce to designate slaves as inferior and as property to be owned and sold as animals were. There was no single event or war-crimes tribunal to end this biopolitical abjection and its impact on those whose bodies were research objects. Instead, decades of civil-rights activism were necessary even to begin to formulate bioethical antidotes to biopolitical abjection of the plantation. Medical experimentation on the bodies of slaves, immigrants, Native Americans, prisoners, and the poor was common, and health disparities reflecting legacies of racial and other axes of discrimination persist.[63] African Americans and other minorities fought, as did women's-health activists, to be self-subjects (as opposed to objects, or subjects substituted for the bodies of others) of research, so as to compile information with which to improve their own health.[64] The Belmont Report began a still far from complete process of research enfranchisement of African American and other minority and low-income populations through amending bioethical principles to reflect that idea that no groups of people should be disproportionately tested on who are unlikely to benefit from the fruits of that biomedical research.

The animal-rights movement makes deeply troubling references to and putative parallels with slavery, thereby perpetuating the "animality" of slavery itself, and simultaneously demeaning the civil-rights struggle by comparing the animal-rights movement to it, though concerns about animal welfare were not absent in civil rights activism.[65] (The same thing occurs when anti-abortion activists claim that abortion, or embryo destruction in stem cell research is similar to slavery.) The historical relations between self-determination and connections to the land and to animals taught by Native American scholars show some ways to think about the powerful as well as the powerless aspects of being treated like animals.[66] Postcolonial theorists have extended this discussion to colonialism, thereby territorializing the biopolitical abject differently, as the world system of metropole and periphery:

Africa is almost always deployed in the framework (or on the fringes) of a meta-text about the *animal*. . . . Is not Africa to be understood for what it is, an entity with its peculiar feature that of shared roots with absolute brutality, sexual license, and death? . . . We can give an account of him / her in the same way we can understand the psychic life of the *beast*. We can even, through a process of domestication and training, bring the African to where he or she can enjoy a fully human life. In this perspective, Africa is essentially, for us, an object of experimentation.[67]

The link between animality and abject poverty and medicine has also been explored.[68] The additions to the Helsinki Declaration that recognize North-South inequity and attempts to protect massive biomedical exploitation of the poor by the rich reflect these kinds of biopolitical abjection.

The successive enfranchisement version of animal rights, which sees animals as the next in line in an essentially Enlightenment project to attribute subjecthood to new classes of hitherto not quite humans (the poor, women, minorities, immigrants, the vanquished, the disabled, and so on) ignores the role that the equation with animality has played historically in denying universal suffrage and justifying abjection. Thinking about animal politics helps to focus on the general problem—for animal and human subjects—of bringing into being the non-substitutional research subject. This is opposed to being either an object of research or a substitutional subject who is tested on largely for the benefit of those human beings whose subjectivity is beyond question. Moving away from substitutional research logic will take a concerted multidisciplinary effort that does not go back on either the scientific or ethical gains that initiated it, but which avoids the scientific and ethical limitations of "like but different" implied by the research model.

THE POLITICS OF ANIMAL RESEARCH

In August 2008, a week before my foray to Googleplex (recounted in chapter 5), and a few months before the incipient meltdown of the California economy and the University of California system had set in, Mark Yudof, the president of the University of California system, issued a statement condemning the setting off of firebombs at UC Santa Cruz by animal-rights activists protesting the use of animals in campus research. In his statement, Yudof referred to "these acts of domestic terrorism."[69] An earlier statement issued by all of the University of California's chancellors referred to "criminal behavior" on the part of animal-rights activists, but did not use the language of terrorism.[70] A few days later, when UC Berkeley's chancellor, Robert Birgeneau, issued his own statement, the language had changed again. The message's title was "Researchers must be free from threats and violence by animal rights terrorists."[71] The *acts* were no longer being condemned as criminal, or as domestic terrorism. Instead, the perpetrators themselves had become "animal rights terrorists." The chancellors, of course, had a duty to protect the research done on their campuses and a fundamental obligation to secure the safety of their faculties, staffs, and students. They also pointed out that animal research has a long history, that has resulted in biomedical breakthroughs that have contributed to global well-being, that the motivations of those conducting animal research are to improve the lot of humanity, and that all campuses adhere strictly to federal and local ethical and regulatory standards for housing, care, and use of animals in research. All in all, it is hard to see how the leaders of these academic institutions could have reacted differently. They had to counter the vilification of researchers and the targeting of their homes, property, and families by animal-rights activists, and they could not afford to compromise on safety. They inherited a long history in the U.S, and elsewhere, of guarding students, faculty members, and facilities against violence perpetrated by animal-rights activists; they were also by no means the first to use the language of terrorism in reference to coordinated animal-rights-related violence, echoing words used by the Federal Bureau of Investigation and in popular discourse. Some different kind of response, particularly any kind of conversation between the sides, would have been close to impossible. It also would have been hard when private capital, government funding, and the research university were colluding to produce a series of elisions: ethics was philanthropy was excess capital was donor funding was research was innovation. Anything that slowed innovation, therefore, was unethical. By 2008, such boom-time elisions had released their hold on the research university to an extent, but not enough to generate a different response from university administrators.

What should be made of the use of terrorism language in reference to animal-rights protesters' attacks on research and researchers?[72] Rethinking the participation of animals in research requires destabilizing the well-rehearsed "animal rights terrorists'" call and response described above. When senior University of California administrators responded to the acts of violence against scientists by using the language of terrorism, they had already placed the university on one side of the violent stalemate needing reconstruction. Other sites within the university could have taken up the charge, but, with research temporarily synonymized with innovation, the kind of archival and contemplative and multi-vocal enterprise that might have opened up the debate was invisible *as* research.

AB 2296, the Researcher Protection Act of 2008, sponsored by the University of California, authored by Gene Mullin (D–South San Francisco) and signed by the governor of California on September 29, 2008, created a new misdemeanor law making it a violation to publish any information about a researcher or his or her family with the intent to aid in committing or threatening violence against the researcher or his or her family. It also created a new misdemeanor trespass that is committed if an animal-rights activist enters onto the residential real property of an academic researcher with the intent to disrupt his or her research. Bringing animal-rights activists and researchers together largely remains to be done.

FROM HUMANIZING THE ANIMAL MODEL TO *IN-VIVO*-IZING THE
IN VITRO MODEL

As the ISSCR 2008 Guidelines for the Clinical Translation of Stem Cells reminded us, "international codes of research ethics, such as the Declaration of Helsinki and the Nuremberg code, strongly encourage the performance of preclinical animal studies prior to clinical trials in humans." This thoughtful document came out in strong favor of both large-animal and small-animal testing of cellular therapies, despite scientific and ethical hurdles. For example, it saw as a shortcoming of large-animal models for cellular therapy testing that "while immunocompromised small-animal models are available for testing, large animals need adjunctive immunosuppressive drug therapy to accept human cell transplants. The side effects of the drugs may interfere with any long-term evaluation of experimental success." Similarly, cell therapy in general poses scientific problems: "It should be acknowledged, however, that preclinical assays including studies in animal models may provide limited insight into how transplanted human cells will behave in human recipients due to the context-dependent nature of cell behavior and the recipient's immune response." In other words, there are special reasons with cellular therapies why results may not be the same across species. In, say, molecular biological processes, there is a high degree of conservation between even simple and complex organisms. Cellular therapies face completely different issues. To deal with the ethical issues, the ISSCR Guidelines recommend special attention to the three R's, reducing numbers of animals used, replacing them with non-animal models where possible, and refining protocols, but they do not advocate against animal testing.

After I had attended many scientific meetings and read countless papers on stem cell research, it became apparent to me that because of certain properties of stem cells, if stem cell research were combined with advances in bioengineering and dynamic visualization techniques, it had potential to be a site where science and ethics might be served by a move away from animal models. University centers were increasingly working on dynamic three-dimensional *in vitro* systems because of the potential of *in vitro* models to improve animal models. According to a 2009 report in the *Boston Globe*,

The feud between animal rights activists and researchers is among the bitterest in science. But many researchers—although adamant that animal research remains critical to finding cures and expanding medical knowledge—have come to concede that using creatures as human stand-ins is unnecessary for many procedures. Indeed, it often isn't even the best science: New drugs that show great promise in mice, for example, often confer zero benefit to humans, or even prove harmful.[73]

At the end of June 2009, Geron announced that it had reached an agreement with GE Healthcare (the company to which Stephen Minger moved—a British subsidiary of General Electric) to "commercialize stem cell drug discovery technologies."[74] As Minger told an interviewer for *Nature*, the aim of the agreement was to conduct drug toxicity trials using *in vitro* human embryonic stem cells instead of laboratory rats, the goal being to improve the capacity to produce hESC lines that are standardized enough to replace animal models. The companies would then be able to produce commercial standardized cellular-assay products that could be used in drug discovery. This had the potential to link "big pharma" with human embryonic stem cell research, and thereby provide a business model, in a near vacuum of such for human cellular technologies and therapies, where they could actually be commercially viable.

In a "backgrounder" document that was available for downloading, Geron spelled out the problems associated with using animals as surrogates for humans, then described how stem cells could start to pave the pathway to human *in vitro* surrogates for humans. It was notable, though, that neither in the announcement of the agreement nor in the background paper was the ethical dimension of reducing animal testing mentioned, but the potential to use the cells instead of animals (and the irony of substituting ethically problematic embryos for scientifically problematic animals) was the focus of a number of non-business news stories.[75]

Here is an excerpt from Geron's document:

Currently, animal models, primary human tissue, and cell lines are used to assess drug metabolism and toxicities. However, these systems have certain limitations. Animal models have an important role in drug metabolism and toxicity studies, but they are not fully reliable predictors of human responses because of basic physiological differences between species. It is not uncommon for the development of a drug to be halted during clinical trials because animal systems did not predict the drug's metabolism or toxicity in humans. A human *in vitro* system commonly used is fresh primary human liver tissue and cells, but access is very limited and the tissue can be variable depending on the donor or the methods used in processing or culturing the samples. Transformed human cell lines have been generated to address supply or variability, but the lines available today do not have the same attributes as their normal counterparts in the body. . . . In contrast, fully functional cells manufactured in bulk from hESCs could be a reliable, uniform and predictive new tool for pharmaceutical companies to perform in vitro metabolism, bidistribution, drug-drug interaction and toxicity testing of drug development candidates.[76]

It is worth noting that the human model is not a straightforwardly good model for humans either, and this applies to cell lines and tissue as much as to clinical trials populations. It takes work to make *any* model system stand in for any target group; there is no mimesis or self-evidently self-same model system. The rise of personalized medicine should, at the very least, rid us of the fantasy that mimesis and servitude are both possible at the same time: if a research subject is being tested on for the benefit of someone else, there are likely to be limits to how good a model the research subject is for those for whom the subject stands in. The more we conceptualize health in terms of genetic and epigenetic and environmental specificity, the less compelling the idea of allowing some to be researched upon for the benefits of others becomes. Instead, a systematic effort at funding, instrumentation, bioengineering, molecular and cell biology, personalized medicine, biobanking, bioethics, and research administration must be undertaken, perhaps overseen by the NIH in conjunction with the major scientific journals and research bodies. We need to begin to work as hard and with as much ingenuity to make life-like *in vitro* human models as scientists and others have worked over the last 60 years to establish the current standards of animal care and use and the extraordinarily refined animal tools and animal-model systems.

IN VITRO VERITAS: THE BEGINNING OF THE END OF THE HUMAN AND ANIMAL RESEARCH SUBJECT

Shiv Visvanathan uses the concepts of triage and vivisection in his account of science, technology, development, and violence in his essay "On the Annals of the Laboratory State." He tells us that "vivisection, which has acquired a central and permanent status within science, has now become totally banal. The pervasive everydayness of it masks the metaphysical shock one would otherwise have experienced. Today, over a hundred million animals are used up in the pursuit of research, in experiments ranging from hair dyes to cancer research. . . . One can add to it now a roll call of patients, prisoners, the poor, inmates of old people's homes, and nameless peasants in the third world." "Opposition to vivisection," Visvanathan argues, "has usually been dismissed by scientists as sentimentalist. But one must see it as a paradigm for general scientific activity extending towards wider domains of control, incorporating innumerable acts of violence within the genre of vivisection."[77] And "triage has been the silent term mediating between the ideas of vivisection and progress."[78] At the end of the beginning of human embryonic stem cell research, it is time to begin to move away from, to begin the end of the acceptability of, the substitutional research subject, human or animal.

As I have argued, not only could many research animals be decommissioned; the science itself promises to be better. Human *in vitro* models could provide a single platform capable of meeting research, development, testing, and therapeutic goals and standards as we move into the era of cell-based therapies. In my opinion, extremism by fringe animal-rights activists should be condemned. Nonetheless, it is possible to respect the views of a sizeable proportion of the population who would like to see the amount of animal testing reduced, regardless of one's one views on the matter. Respecting views means taking them into account in a meaningful way. It does not mean agreeing with them. That is the essence of pluralism, and democracies fare better when widespread moral concerns are respected. No researcher or administrator has to agree with proponents of animal rights to prefer, where possible, to take their views into consideration. To open up this position, it is necessary first of all to see that science and ethics are not opposed on this issue, and that, on the contrary, ethical concern lies at the heart of innovation. Many scientists, if not most, enter the field to make the world better. This effort would be entirely consistent with that enterprise.

Remarkable progress has been made toward humanizing animal models in recent years, especially with the rise of stem cell research. Systematic efforts should be made to improve *in vitro* human models in such as way as to make them more *in vivo*-like, rather than putting resources and energy into humanizing non-human animal models. For example, combining bioengineering efforts to make synthetic biosafe tissue and organ models, biophysics advances in dynamic visualization, molecular and cell biology advances in knowledge of transcription factors that organize differentiation, and cellular biological knowledge of transplanted cell migration, division, and determination have the potential together to improve upon animal models. A change from humanization of model systems to *in-vivo*-ization of *in vitro* systems also has the potential to have a human distributive justice payoff. Instead, a human *in vivo* model would be customized with a patient's own cells or genomic information, making it much less relevant how a therapy behaves in a generic human, and thus making it increasingly redundant to have general human clinical trials. This, is turn, would help cut out the step whereby one group of humans with greater economic need (including the phenomenon of "clinical trials outsourcing" to poorer people or people with lesser freedom (e.g., prisoners) become biological test sites for drugs and therapies that are often inaccessible to them or are aimed at treating diseases of the affluent. Good science demands no less.

Pluripotent stem cell research faces a number of challenges in the coming decades and holds tremendous promise. Regulations and ethical guidelines that reflect certain values (some of them national and some of them religious) have been implemented in many places around the world. Efforts are well underway to harmonize and standardize regulations between jurisdictions enough so that researchers and materials and knowledge can be shared among different regions and nations. Despite (or perhaps because of) the visibility of its early challenges, stem cell research has the chance to become a vibrant scientific, medical, and biotechnological field that could also be among the best-regulated areas of science the modern world has seen. With an effort to foster the "good science" called for in this book, questions at the confluence of science and ethics could begin to effect a change in biopolitical paradigm. But if the ethical momentum is lost as these technologies normalize, they may fall into all-too-familiar patterns of biomedical research, replicating R&D and market practices that were not developed for human tissue economies and exacerbating inequalities in health care. The stakes are high, and the time is short.

Tissue engineers, reproductive and cell biologists, geneticists, clinicians, social scientists, lawyers, administrators, and others are just beginning to understand what it takes to work reliably with these fundamental building blocks of the human body. Good stem cell science will require satisfying public concerns about the safety and efficacy of stem-cell-based therapies while moving the science from the laboratory to clinical trials and on to treatment. Scientists and physicians, and disease advocates and other enthusiasts, will have to communicate clearly with the rest of the public, differentiating important but not necessarily empirically accurate forward-looking rhetoric from the social and scientific processes actually underway. In the United States, many investors and corporations remain to be convinced that it is a good financial bet to bring pluripotent-stem-cell-based therapies to the market, in view of the untested nature of public financing for a tissue infrastructure and the paucity of business models for human biologicals. The private sector needs to convince people that public gifts of money, tissue, and bioinformation are not a property grab, and that they support a new knowledge and practice for the common good. Issues about the provenance, procurement, and disposition of pluripotent stem cell lines and their raw materials and products must be expected to be ongoing, and the potential for disagreement and the consequent need for changes in regulation and practice should be honored by making regulations transparent, democratically governed, revisable, and flexible to the greatest extent possible.

Stem cell research, especially publicly funded research, must not be allowed to exacerbate current disparities in health. To avoid ushering in a culture of tissue screening and embryo screening, opinions from disability justice spokespeople and disability scholarship should be taken into account every step of the way. Though it is important to focus on quality control in stem cell tourism, it is equally important to consider what makes some of the world's most privileged people undertake medical travel and some of the world's least privileged people commit their bodies to the treatment of the wealthy. As biobanks and bioinformatics grow rapidly around the world, it will be important to try to avoid a schism where some people have their biological material and data stored for personalized medicine, while others have theirs stored in government law enforcement or surveillance facilities; one way to avoid this would be for tissue and information to be stored and shared in ways where it could be used for personalized medicine as easily as for forensic or government surveillance purposes and vice versa. This would require considerable effort to overcome privacy and secrecy concerns about data sharing and to overcome the difficulties of moving between the private and public sectors, but the potential benefit to health equity would be significant. Collaborations involving tissue engineers, molecular and cell biologists, and biophysics visualization specialists should be actively encouraged to improve *in vitro* models of human biological systems, with an explicit aim of decommissioning some of the humans and some of the animals currently deployed as research subjects. Good science cannot be achieved or legislated once and for all. It is ongoing and reiterative, and it requires openness to dissent and the best work of many different kinds of contributors. But as our bodies become not only ourselves, but the Other (and vice versa), the opportunity and challenge for co-produced ethics and science has rarely been greater.

APPENDIX A   GLOSSARY

ANIMAL MODEL

A laboratory animal that is used to express symptoms of a human disease for purposes such as verifying a scientific principle, testing experimental medical treatments, or drug discovery.

ASSAY (OR BIOASSAY)

A determination of the strength or biological activity of a substance, such as a drug, by comparing its effects with those of a standard preparation on a test organism, or a test used to determine such a strength or activity.

BIOBANKS (OR BIOREPOSITORIES)

Biobanks integrate collections of bio-specimens (cell lines, including stem cells, blood, DNA, tissue, and biopsy specimens) with corresponding data on patients, such as genetic profiles, medical histories, and lifestyle information. They represent the continuing convergence of the life and information sciences, and are a new frontier for regenerative medicine and genomics.

BIOLOGICAL (NOUN)

A preparation, such as a drug, a vaccine, or an antitoxin, that is synthesized from living organisms or their products and used medically as a diagnostic, preventive, or therapeutic agent. Human biologicals are derived from humans.

BLASTOCYST

A pre-implantation embryo of 30–150 cells. The blastocyst consists of a sphere made up of an outer layer of cells (the trophectoderm), a fluid-filled cavity (the blastocoel), and a cluster of cells on the interior (the inner cell mass, from where stem cells may be derived). A blastocyst is approximately 0.15 millimeters in diameter.

CELL THERAPY

The administration of genetically engineered cells, healthy donor cells, or a patient's own stem cells for the purpose of medical treatment.

CHIMERA

A human embryo consisting of cells from more than one genetically distinct human embryo, fetus, human being, or non-human life form.

CLONING

The intentional creation of one or more genetically identical embryos for either reproductive or therapeutic purposes. "Reproductive cloning" refers to the cloning of an embryo for transplantation into a uterus with the intention of producing offspring genetically identical to the nuclear donor. "Therapeutic cloning" or "research cloning" refers to the cloning of an embryo for the purpose of deriving stem cells for therapeutic uses.

DRUG DISCOVERY

The process by which drugs are discovered and/or designed. The research process usually identifies molecules with the desired biological effects in animal models.

GENOMICS

The systematic study of the structure of the genome of an organism, including mapping and sequencing.

GERM (SEX) CELL

A sperm cell, an egg cell, or a precursor of one. Each mature germ cell is haploid— that is, it has a single set of 23 chromosomes, containing half the usual amount of DNA. Other body cells are called somatic cells and are diploid, containing 46 chromosomes.

GERM-LINE GENETIC MODIFICATION (OR GERM-LINE ALTERATION)

The modification of a human genome such that the modification is passed on to descendants.

HEMATOPOIETIC STEM CELLS (HSCS)

Multipotent stem cells that give rise to all types of blood cells, including myeloid (monocytes and macrophages, neutrophils, basophils, eosinophils, erythrocytes, megakaryocytes/platelets, dendritic cells) and lymphoid lineages (T cells, B cells, NK cells).

HISTOCOMPATIBILITY

The property of a donor of tissue, cells, or blood, and a recipient having the same, or mostly the same, alleles of a set of genes called the major histocompatibility complex. These genes are expressed in most tissues as antigens, to which the immune system makes antibodies. The immune system at first makes antibodies to all sorts of antigens, including those it has never been exposed to, but stops making them to antigens present in the body. If the body is exposed to foreign antigens, as by getting a tissue graft, it attacks the foreign material unless it is histocompatible.

HUMAN EMBRYONIC STEM CELLS

Cells found in a blastocyst (a four-to-five-day post-fertilization embryo), which can be extracted from the embryo and induced to divide, producing a pluripotent cell line, capable of differentiating into the different cell lineages of the human body. Most human embryonic stem cells are derived from embryos that develop from eggs that have been fertilized in an *in vitro* fertilization clinic and then donated, with the informed consent of the donors, for research purposes.

HYBRID

An organism whose cells contain DNA from different species. In humans, an ovum that has been fertilized by sperm from a different species or into which the nucleus or cytoplasm of a cell of a non-human life form has been placed or vice versa.

INDUCED PLURIPOTENT STEM CELL (IPS CELL, OR IPSC)

A pluripotent stem cell derived from a non-pluripotent cell (typically an adult somatic cell) by reverse engineering the cell and inducing the expression of certain genes.

INFORMED CONSENT

Written consent, obtained after information has been received and understood by a person, stating that the person agrees to undergo a medical procedure or to participate in a clinical trial.

IN VITRO

(Literally, in glass.) Outside the body—for example, in a Petri dish.

IN VIVO

Within the body—for example, an *in vivo* embryo is inside a woman, in her uterus.

LEFTOVER (SUPERNUMERARY, SPARE) EMBRYO

An embryo—either fresh or frozen—that was derived during, but not used in and no longer wanted for, assisted reproductive technology treatment for infertility or disease deselection.

MEDICAL TOURISM
A term used by travel agencies, governments, and the mass media to describe the rapidly growing practice of traveling, often across international borders, to obtain health care.

PLURIPOTENT
Pertaining to the ability of a cell to differentiate into many cell types. Examples of pluripotent cells are stem cells in animals and meristematic cells in higher plants. Pluripotent stem cells are slightly different from totipotent cells; totipotent stem cells can produce all the cells of the body. Pluripotent stem cells can differentiate into all cell types except those that form the trophoblast. Multipotent stem cells, such as hematopoietic stem cells, can differentiate into several different tissue types.

PRE-IMPLANTATION GENETIC DIAGNOSIS (PGD)
A technique by which *in vitro* embryos are tested for specific genetic disorders (e.g., cystic fibrosis) or other characteristics (e.g., sex) before being transferred to a woman's uterus. Usually used to prevent passing on a serious disease, but also used for tissue matching for "savior siblings," and with the potential to be used for trait selection.

REGENERATIVE MEDICINE
The study and laboratory and clinical development of replacement cells, tissues, and organs, to regenerate injured or diseased tissue, and to increase longevity.

STEM CELL
An unspecialized cell characterized by the ability to self-renew by mitosis while in an undifferentiated state and having the capacity to give rise to various differentiated cell types upon differentiation.

TISSUE ENGINEERING
A field of inquiry and research that combines cell biology, engineering, and biochemistry to help replace, repair, or re-grow injured or diseased tissues. An important part of regenerative medicine and of synthetic biology.

TRIAGE
A process for sorting injured people into groups on the basis of their respective needs for or likely capacity to benefit from immediate medical treatment. Triage is used in hospital emergency rooms, on battlefields, and at disaster sites when limited medical resources must be allocated. More generally it can be used to refer to any process in which things are ranked in terms of importance or priority for the purposes of assigning timing and type of treatment.

The definitions above were adapted from the following sources:

The American Heritage Dictionary of the English Language, fourth edition, copyright 2000 by Houghton Mifflin Company. Updated in 2009. Published by Houghton Mifflin Company. (assay, triage, biologicals)
Biology Online Dictionary (http://www.biology-online.org/dictionary) (stem cells, pluripotent)
Clinfowiki (http://www.clinfowiki.org/wiki) (biobanking)
Health Canada (http://www.hc-sc.gc.ca/hl-vs/reprod/hc-sc/gloss/index -eng.php) (blastocyst, chimera, cloning, genomics, germ cell, germ-line alteration, hybrid, informed consent, pre-implantation genetic diagnosis, supernumerary embryo)
McGowan Institute for Regenerative Medicine and UPMC Health System (http://www.upmc.com/SERVICES/MIRM/REGENERATIVEMEDI-CINE/Pages/glossary-terms.aspx) (regenerative medicine, animal model, cell therapy, tissue engineering)
National Institutes of Health (http://stemcells.nih.gov/info/basics/basics3 .asp) (human embryonic stem cell research)
University of California San Francisco Pharmacy (http://pharmacy.ucsf .edu/glossary/d/) (drug discovery)
Wikipedia (http://en.wikipedia.org) (medical tourism, induced pluripotent stem cells, drug discovery, histocompatibility)

APPENDIX B   RESOURCES AND PRIMARY DOCUMENTS FOR STEM CELL RESEARCH INVOLVING EMBRYO(ID) POTENTIAL SUBJECTS, AND SELECTED TEXT

GENERAL RESOURCES

CALIFORNIA
California Institute for Regenerative Medicine, http://www.cirm.ca.gov/
California Department of Public Health Human Stem Cell Research Program, http://www.cdph.ca.gov/

U.S. FEDERAL
NIH Stem Cell Information, U.S., http://stemcells.nih.gov/
Presidential Commission for the Study of Bioethical Issues, U.S., http://bioethics.gov/
Stem Cells at the National Academies, U.S., http://dels-old.nas.edu/bls/stemcells/
Interstate Alliance on Stem Cell Research, http://www.iascr.org/
StemCellResources.org, http://www.stemcellresources.org/

INTERNATIONAL/OTHER COUNTRIES
International Society for Stem Cell Research, http://www.isscr.org/
International Stem Cell Forum, http://www.stem-cell-forum.net/ISCF/
Stem Cell Society, Singapore, http://www.stemcell.org.sg/
SNAP: Stem Cell Network Asia Pacific, http://www.asiapacificstemcells.org/
The Hinxton Group, http://www.hinxtongroup.org/
StemGen, http://www.stemgen.org/
Human Embryology and Fertilisation Authority, United Kingdom, http://www.hfea.gov.uk/
EuroStemCell, http://www.eurostemcell.org/

ADVOCACY IN FAVOR OF PERMITTING AND FUNDING ALL STEM
CELL RESEARCH, ACTIVE IN CALIFORNIA AND U.S. NATIONAL
STEM CELL DEBATE

Research America: An Alliance for Discoveries in Health (pro-research),
http://www.researchamerica.org/stemcell_issue
Christopher and Dana Reeve Foundation (patient advocacy), http://www
.christopherreeve.org/
Americans for Cures (patient advocacy), http://www.americansforcures
.org/

ADVOCACY IN FAVOR OF RESTRICTING OR REGULATING HUMAN
EMBRYONIC AND CERTAIN OTHER KINDS OF STEM CELL
RESEARCH, ACTIVE IN CALIFORNIA AND U.S. NATIONAL STEM
CELL DEBATE

Center for Genetics and Society, California, U.S. (progressive civil society
group in favor of broader regulation of stem cell research), http://www
.geneticsandsociety.org/
Generations Ahead, California, U.S. (fights for race, gender, disability social
justice in repro-genetics; closed in January 2012 but website remains
active), http://www.generations-ahead.org/
Nightlife Christian Adoptions (runs Snowflakes Embryo Adoption program),
http://www.nightlight.org/apply-for-adoption/snowflakes-embryo.aspx
California Family Council (promotes right-wing evangelical Christian
policy in California), http://www.californiafamilycouncil.org/about-us
Life Legal Defense Fund, CA (documents and promotes pro-life legal
initiatives in California), http://lldf.org/stemcell/

PRIMARY TEXTS

Proposition 71, California, 2004 (available at http://www.cirm.ca.gov/
sites/default/files/files/about_cirm/prop71.pdf)
California Code of Regulations, Title 17 Public Health, Division 4 California Institute for Regenerative Medicine, California, 2010
Dickey-Wicker Amendment, 2009 (text available at http://www.govtrack
.us/congress/bills/111/hr1105/text)
National Institutes of Health Guidelines on Human Stem Cell Research,
74 Fed. Reg. 32,170–32,175 (July 7, 2009)

Final Report of the National Academies' Human Embryonic Stem Cell Research Advisory Committee, with Amendments to the National Academies' Guidelines for Human Embryonic Stem Cell Research, 2010
California Department of Public Health Guidelines for Human Stem Cell Research, 2009.
Embryo donation for stem cell research (collected for fertility purposes and in excess of clinical need), ISSCR sample consent document, South Korea's Bioethics and Safety Act, 2008.
Singapore's Human Cloning and Other Prohibited Practices Act, 2004
Doe v. Obama, United States Court of Appeal for the Fourth Circuit, Argued December 7, 2010; Decided January 21, 2011
Sherley v. Sibelius, United States Court of Appeals for the District of Columbia Circuit, Argued December 6, 2010; Decided April 29, 2011
Evangelium vitae on the Value and Inviolability of Human Life, Pope John Paul II, The Vatican, 1995

SELECTED TEXTS

AMENDED "ACCEPTABLE RESEARCH MATERIALS" AND "USE OF FETAL TISSUE" TO QUALIFY FOR CALIFORNIA INSTITUTE FOR REGENERATIVE MEDICINE FUNDING, CALIFORNIA CODE OF REGULATIONS
Eff. Dec., 2010 1 OAL Approved
Amend 17 Cal. Code of Regs. section 100080 to read:
§ 100080. Acceptable Research Materials.
All covered stem cell lines used in CIRM-funded research must be "acceptably derived."

(a) To be "acceptably derived," the covered stem cell line must meet one of the following three criteria:

   (1) The covered stem cell line is recognized by an authorized authority . . .
   (2) The covered stem cell line is derived under the following conditions:

      (A) Donors of human gametes, embryos, somatic cells or tissue gave voluntary and informed consent; and

(B) Donors of human gametes or embryos did not receive valuable consideration. For embryos originally created using in vitro fertilization for reproductive purposes and are no longer needed for this purpose, "valuable consideration" does not include payments to original gamete donors in excess of "permissible expenses." . . .

(C) Donation of human gametes, embryos, somatic cells or tissue was overseen by an IRB (or, in the case of foreign sources, an IRB-equivalent); and

(D) Individuals who consented to donate stored human gametes, embryos, somatic cells or tissue were not reimbursed for the cost of storage prior to donation.

(3) The covered stem cell line is derived from non-identifiable human somatic cells under the following conditions . . .

(A) The derivation did not result from the transfer of a somatic cell nucleus into a human oocyte (SCNT) or the creation or use of a human embryo; and

(B) The somatic cells have no associated codes or links maintained by anyone that would identify to the investigator(s) the donor of the specimens . . .

(b) In addition to the requirements of subdivision (a) of this chapter, the following requirements apply to the derivation and use of all covered stem cell lines.

(1) Any covered stem cell line derived from any intact human embryo, any product of SCNT, parthenogenesis or androgenesis after 12 days in culture may not be used unless prior approval is obtained from the Independent Citizens Oversight Committee, constituted under Health & Safety Code, section 125290.15. Use of any covered stem cell line derived from any intact human embryo, any product of SCNT, parthenogenesis or androgenesis after 14 days or after the appearance of the primitive streak is prohibited. The 12–14 day limit does not include any time during which the cells have been frozen.

(2) Any payments for the purchase of covered stem cell lines, somatic cells, or human tissue to persons other than the original donors shall be limited to those costs identified in Health & Safety Code, section 125290.35, subdivision (b)(5). Any payment for gametes and embryos, to persons other than the original donors, shall be limited to necessary and reasonable costs directly incurred as a result of providing materials for research, which include but are not limited to expenditures associated with processing, quality control, storage, or transportation. Note: Authority cited: Article XXXV, California Constitution; Section 125290.40(j), Health and Safety Code. Reference: Sections 125290.35, 125290.40, 125290.55 and 125300, Health and Safety Code.

Eff: 06/29/08 1 100085 AMENDED OAL APPROVED
Amend 17 Cal. Code of Regs. section 100085 to read:
§ 100085. Use of Fetal Tissue.
Fetal tissue shall be procured in accordance with 17 Cal. Code Regs. section 100080, subdivision (a)(2). In addition, research involving human fetal tissue will adhere to the following provisions:

(a) The woman who donates the fetal tissue must sign a statement declaring:

(1) That the donation is being made for research purposes, and
(2) The donation is made without any restriction regarding who may be the recipient(s) of materials derived from the tissue; and

(b) The attending physician must:

(1) Sign a statement that he or she has obtained the tissue in accordance with the donor's signed statement. In the case of tissue obtained pursuant to an induced abortion, the physician must sign a statement stating that he or she:

(A) Obtained the woman's consent for the abortion before requesting or obtaining consent for the tissue to be used for research;

(B) Did not alter the timing, method, or procedures used to terminate the pregnancy solely for the purpose of obtaining the tissue for research; and

(C) Performed the abortion in accordance with applicable law.

(2) Disclose to the donor any financial interest that the attending physician has in the research to be conducted with the tissue.

(3) Disclose any known medical risks to the donor or risks to her privacy that might be associated with the donation of the tissue and that are in addition to risks of such type that are associated with the woman's medical care.

(c) The principal investigator of the research project must sign a statement certifying that he or she:

(1) Is aware that the tissue is human fetal tissue obtained in a spontaneous or induced abortion or pursuant to a stillbirth;

(2) Is aware that the tissue was donated for research purposes;

(3) Had no part in any decisions as to the timing, method, or procedures used to terminate the pregnancy; and

(4) Is not the donor's attending physician.

Note: Authority cited: California Constitution, article XXXV; Section 125290.40(j), Health and Safety Code. Reference: Sections 125290.35, 125290.40, 125290.55 and 125300, Health and Safety Code.

(ii) Parties to *Doe v. Obama*
MARY SCOTT DOE, a human embryo "born" in the United States (and subsequently frozen in which state of cryopreservation her life is presently suspended), individually and on behalf of all other frozen human embryos similarly situated; NATIONAL ORGANIZATION FOR EMBRYONIC LAW (NOEL); NIGHTLIGHT CHRISTIAN ADOPTIONS; PETER MURRAY; SUZANNE MURRAY; COURTNEY ATNIP; TIM ATNIP;

STEVEN B. JOHNSON; KATE ELIZABETH JOHNSON; CORA
BEST; GREGORY BEST, Plaintiffs-Appellants, v.
BARACK HUSSEIN OBAMA, in his official capacity as President of the
United States; CHARLES E. JOHNSON, in his official capacity as acting
secretary of the Department of Health & Human Services; RAYNARD
S. KINGTON, in his official capacity as acting director of the National
Institutes of Health, Defendants-Appellees.
(iii) non-applicability of Dickey-Wicker, NIH Guidelines on Human Stem
Cell Research, 2009

Since 1999, the Department of Health and Human Services (HHS) has
consistently interpreted this provision as not applicable to research using
hESCs, because hESCs are not embryos as defined by Section 509. This
long-standing interpretation has been left unchanged by Congress, which
has annually reenacted the Dickey Amendment with full knowledge that
HHS has been funding hESC research since 2001. These guidelines therefore recognize the distinction, accepted by Congress, between the derivation of stem cells from an embryo that results in the embryo's destruction,
for which federal funding is prohibited, and research involving hESCs that
does not involve an embryo nor result in an embryo's destruction, for
which federal funding is permitted.

APPENDIX C   RESOURCES AND PRIMARY DOCUMENTS FOR STEM CELL RESEARCH INVOLVING HUMAN SUBJECTS, AND SELECTED TEXTS

GENERAL RESOURCES

Office for Human Research Protections (OHRP), U.S. Department of Health and Human Services, http://www.hhs.gov/ohrp/
Office of Human Subjects Research, National Institutes of Health, U.S., http://ohsr.od.nih.gov/
Presidential Commission for the Study of Bioethical Issues, http://www.bioethics.gov/
Food and Drug Administration (FDA), U.S. Department of Health and Human Services, http://www.fda.gov/
World Medical Association, http://www.wma.net/
UNESCO Bioethics Programme, http://www.unesco.org/new/en/social-and-human-sciences/themes/bioethics/

PRIMARY TEXTS

The Nuremberg Code: Directives for Human Experimentation, 1947
World Medical Association Declaration of Helsinki: Ethical Principles for Medical Research Involving Human Subjects, with amendments up to and including WMA General Assembly, 1964 / Tokyo 2004
Code of Federal Regulations Title 45 Volume 46 (45 CFR 46), the "Common Rule," including revisions as of June 23, 2005
Advanced Notice of Proposed Rule Making (ANPRM) for Revisions to the Common Rule, July 22, 2011
The Belmont Report: Ethical Principles and Guidelines for the protection of human subjects of research, National Commission for the Protection of Human Subjects of Biomedical and Behavioral Research, April 18, 1979.

Senate Bill 1260. Senator Deborah Ortiz. Reproductive Health and Research, California, 2006

Medical Risks of Oocyte Donation for Stem Cell Research: Workshop Summary, Institute of Medicine, National Academies, 2007.

Egg donation for stem cell research; provided directly and solely for stem cell research, ISSCR sample consent document

Egg donation for stem cell research; collected during the course of fertility treatment and in excess of clinical need, ISSCR sample consent document

Sperm donation for stem cell research, ISSCR sample consent document

Somatic cell donation for stem cell research, ISSCR sample consent document

Guidelines for the Clinical Translation of Stem Cells, International Society for Stem Cell Research, 2008.

SELECTED TEXT

(i) Amended "Informed Consent Requirements," California Institute for Regenerative Medicine, California Code of Regulations
Eff: 06/29/08 1 100100 AMENDED OAL APPROVED
1 Amend 17 Cal. Code of Regs. section 100100 to read:
§ 100100. Informed Consent Requirements.

(a) All CIRM-funded human subjects research shall be performed in accordance with Title 45 Code of Federal Regulations, Part 46 (Protection of Human Subjects), revised June 23, 2005, and California Health and Safety Code section 24173. In accordance with existing law, California Health and Safety Code section 24173 does not apply to a person who is conducting research as an investigator within an institution that holds an assurance with the United States Department of Health and Human Services pursuant to Title 45 Code of Federal Regulations Part 46, revised June 23, 2005, and who obtains informed consent in the method and manner required by those regulations.

(b) In addition to the requirements of Code of California Regulations, title 17, section 100080, subdivision (a)(2), the following provisions apply when CIRM funded research involves donation of human gametes, embryos, somatic cells or tissue for derivation of new covered stem cell lines:

(1) CIRM-funds may not be used for research that violates the documented preferences of donors with regard to the use of donated materials. The SCRO committee or IRB must confirm that donors have given voluntary and informed consent in accordance with this section. To ensure that donors are fully informed of the potential uses of donated materials in addition to the general requirements for obtaining informed consent identified in subdivision (a) of this regulation, researchers shall disclose all of the following, unless a specific item has been determined by the SCRO committee or IRB to be inapplicable:

   (A) Derived cells or cell products may be kept for many years
   (B) Whether or not the identity(ies) of the donor will be ascertainable by those who work with the resulting cells or cell products. If the identity of the donor is to remain associated with the cells or cell products, then the investigator must inform the donor of any plan for recontact whether for the purpose of providing information about research findings to donors, or for the purpose of requesting additional health information. After donation, an investigator may recontact a donor only if the donor consents at the time of donation.
   (C) Cell lines may be used in future studies which are not now foreseeable.
   (D) Derived cells or cell products may be used in research involving genetic manipulation.
   (E) Derived cells or cell products may be transplanted into humans or animals.
   (F) Derived cells or cell products are not intended to provide direct medical benefit to the donor, except in the case of autologous donation.
   (G) The donation is being made without restriction on the recipient of transplanted cells, except in the case where donation is intended for autologous transplantation.

(H) Neither consent nor refusal to donate materials for research will affect the quality of any care provided to a potential donor.

(I) Although the results of research including donated materials may be patentable or have commercial value, the donor will have no legal or financial interest in any commercial development resulting from the research.

(2) A donor must be given the opportunity to impose restrictions on future uses of donated materials. Researchers may choose to use materials only from donors who agree to all future uses without restriction.

(3) For CIRM-funded research involving the donation of oocytes, an IRB finding that potential risks of donation are reasonable even if there is no anticipated benefit to the donor shall be documented and made available to the donor, SCRO and the CIRM. In addition, the following requirements apply:

(A) The description of foreseeable risk required in subdivision (a) of this regulation shall include but not be limited to information regarding the risks of ovarian hyperstimulation syndrome, bleeding, infection, anesthesia and pregnancy.

(B) Any relationship between the attending physician and the research or researcher(s) must be disclosed to an egg donor.

(C) Prospective donors shall be informed of their option to deliberate before deciding whether or not to give consent. If a deliberation period is chosen, the donor shall be informed of her right to determine the method of recontact. The donor must be informed that she has the option to initiate recontact. Investigators shall not initiate recontact unless the donor has consented, and this consent is documented in the research record.

(D) The researcher shall ascertain that the donor understands the essential aspects of the research involving donated materials, following a process approved by the designated IRB or SCRO committee. Understanding the essential aspects of the research includes understanding at least that:

(i) Eggs will not be used for reproductive purposes.

(ii) There are medical risks in oocyte donation, including the risks of ovarian hyperstimulation syndrome, bleeding, infection, anesthesia, and pregnancy.

(iii) The research is not intended to directly benefit the donor or any other individual.

(iv) Whether stem cell lines will be derived from her oocytes through fertilization, SCNT, parthenogenesis, or some other method.

(v) Stem cell lines developed from her oocytes will be grown in the lab and shared with other researchers for studies in the future.

(vi) If stem cells derived from her donation are to be transplanted into patients, researchers might recontact the donor to get additional health information.

(vii) Donors receive no payment beyond reimbursement for permissible expenses.

(viii) Stem cell lines derived as a result of her oocyte donation may be patented or commercialized, but donors will not share in patent rights or in any revenue or profit from the patents.

(4) For funded research involving the donation and destruction of human embryos for stem cell research, the informed consent process shall include a disclosure that embryos will be destroyed in the process of deriving embryonic stem cells.

(5) Research that uses human umbilical cord, cord blood or placenta, consent shall be obtained from the birth mother.

(6) For research involving the donation of somatic cells for SCNT, the informed consent process shall include disclosure as to whether the donated cells may be available for autologous treatment in the future.

Note: Authority cited: Article XXXV, California Constitution; Section 125290.40(j), Health and Safety Code. Reference: Sections 24173, 125290.35, 125290.40, 125290.55 and 125315, Health

(ii) Stem cell tourism positions statement, Guidelines for the Clinical
    Translation of Stem Cells, ISSCR, 200

2. Position on Unproven Commercial Stem Cell Interventions

The ISSCR recognizes an urgent need to address the problem of unproven
stem cell interventions being marketed directly to patients. Numerous
clinics around the world are exploiting patients' hopes by purporting to
offer new and effective stem cell therapies for seriously ill patients, typically
for large sums of money and without credible scientific rationale, transparency, oversight, or patient protections. The ISSCR is deeply concerned
about the potential physical, psychological, and financial harm to patients
who pursue unproven stem cell-based "therapies" and the general lack of
scientific transparency and professional accountability of those engaged in
these activities.

The marketing of unproven stem cell interventions is especially worrisome in cases where patients with severe diseases or injuries travel across
borders to seek treatments purported to be stem cell-based "therapies" or
"cures" that fall outside the realm of standard medical practice. Patients
seeking medical services abroad may be especially vulnerable because of
insufficient local regulation and oversight of host clinics. Some locales may
further lack a system for medical negligence claims, and there may be less
accountability for the continued care of foreign patients. To help address
some of these concerns, the individuals and their doctors make informed
choices when contemplating a stem cell-based intervention either locally
or abroad.

(iii) selection from California SB 1260. Ortiz. Reproductive Health and
    Research

Section 1 . . .

(c) This act seeks to support the requirements already in current law
    upholding the principle of voluntary and informed consent and to
    tailor them to this new area of pioneering research that utilizes
    human oocytes.

(d) The potential for exploitation of the reproductive capabilities of
    women for commercial gain raises health and ethical concerns that
    justify the prohibition of payment for human oocytes. . . .

125350. No human oocyte or embryo shall be acquired, sold, offered for sale, received, or otherwise transferred for valuable consideration for the purposes of medical research or development of medical therapies. For purposes of this section, "valuable consideration" does not include reasonable payment for the removal, processing, disposal, preservation, quality control, and storage of oocytes or embryos.

125355. No payment in excess of the amount of reimbursement of direct expenses incurred as a result of the procedure shall be made to any subject to encourage her to produce human oocytes for the purposes of medical research.

APPENDIX D   RESOURCES AND PRIMARY DOCUMENTS FOR STEM CELL RESEARCH INVOLVING NON-HUMAN ANIMAL SUBJECTS, AND SELECTED TEXT

GENERAL RESOURCES

Office of Laboratory Animal Welfare (OLAW), http://grants.nih.gov/grants/olaw

American Association for Laboratory Animal Science, http://www.aalas.org/

Altweb: the global clearing house for information on alternatives to animal testing, http://altweb.jhsph.edu/

Animal Research for Life, Europe, http://www.animalresearchforlife.eu/

PRIMARY TEXTS

U.S. Animal Welfare Act, 1966, with amendments through 2008: Public Law 89-544, 1966, as amended, (P.L> 91-579, P.L. 94-279 and P.L. 99-198 7 U.S.C. 2131 et. Seq. Implementing regulations Code of Federal Regulations (CFR), Title 9, Chapter 1, Subchapter A, Parts 1, 2, and 3; administered by U.S. Department of Agriculture

Guide for the Care and Use of Laboratory Animals, 8th Edition, Institute of Laboratory Animal Resources, The National Academy of Science, The National Academies Press, 2010

American Veterinary Medical Association Guidelines on Euthanasia, 2007, revised from *Journal of the American Veterinary Medical Association* 218 (2001), no. 5: 669–696

International Guiding Principles for Biomedical Research Involving Animals, Council for International Organizations of Medical Sciences (CIOMS), 1985

European Convention for the Protection of Vertebrate Animals use for Experimental and Other Scientific Purposes; ETS No. 123, 1986, with amendment from 2005, ETS No. 170.

SELECTED TEXT

INTERNATIONAL GUIDING PRINCIPLES FOR BIOMEDICAL RESEARCH INVOLVING ANIMALS

PREAMBLE   Experimentation with animals has made possible major contributions to biological knowledge and to the welfare of man and animals, particularly in the treatment and prevention of diseases. Many important advances in medical science have had their origins in basic biological research not primarily directed to practical ends as well as from applied research designed to investigate specific medical problems. There is still an urgent need for basic and applied research that will lead to the discovery of methods for the prevention and treatment of diseases for which adequate control methods are not yet available—notably the noncommunicable diseases and the endemic communicable diseases of warm climates.

Past progress has depended, and further progress in the foreseeable future will depend, largely on animal experimentation which, in the broad field of human medicine, is the prelude to experimental trials on human beings of, for example, new therapeutic, prophylactic, or diagnostic substances, devices, or procedures.

There are two international ethical codes intended principally for the guidance of countries or institutions that have not yet formulated their own ethical requirements for human experimentation: The Tokyo revision of the Declaration of Helsinki of the World Medical Association (1975); and the Proposed International Guidelines for Biomedical Research Involving Human Subjects of the Council for International Organizations of Medical Sciences and the World Health Organization (1982). These codes recognize that while experiments involving human subjects are a sine qua non of medical progress, they must be subject to strict ethical requirements. In order to ensure that such ethical requirements are observed, national and institutional ethical codes have also been elaborated with a view to the protection of human subjects involved in biomedical (including behavioral) research.

A major requirement both of national and international ethical codes for human experimentation, and of national legislation in many cases, is that new substances or devices should not be used for the first time on human beings unless previous tests on animals have provided a reasonable presumption of their safety.

The use of animals for predicting the probable effects of procedures on human beings entails responsibility for their welfare. In both human and veterinary medicine animals are used for behavioral, physiological, pathological, toxicological, and therapeutic research and for experimental surgery or surgical training and for testing drugs and biological preparations. The same responsibility toward the experimental animals prevails in all of these cases.

Because of differing legal systems and cultural backgrounds there are varying approaches to the use of animals for research, testing, or training in different countries. Nonetheless, their use should be always in accord with humane practices. The varying approaches in different countries to the use of animals for biomedical purposes, and the lack of relevant legislation or of formal self-regulatory mechanisms in some, point to the need for international guiding principles elaborated as a result of international and interdisciplinary consultations.

The guiding principles proposed here provide a framework for more specific national or institutional provisions. They apply, not only to biomedical research but also to all uses of vertebrate animals for other biomedical purposes, including the production and testing of therapeutic, prophylactic, and diagnostic substances, the diagnosis of infections and intoxications in man and animals, and to any other procedures involving the use of intact live vertebrates.

I . BASIC PRINCIPLES

I. The advancement of biological knowledge and the development of improved means for the protection of the health and well-being both of man and of animals require recourse to experimentation on intact live animals of a wide variety of species.

II. Methods such as mathematical models, computer simulation and in vitro biological systems should be used wherever appropriate.

III. Animal experiments should be undertaken only after due consideration of their relevance for human or animal health and the advancement of biological knowledge.

IV. The animals selected for an experiment should be of an appropriate species and quality, and the minimum number required to obtain scientifically valid results.

V. Investigators and other personnel should never fail to treat animals as sentient, and should regard their proper care and use and the avoidance or minimization of discomfort, distress, or pain as ethical imperatives.

VI. Investigators should assume that procedures that would cause pain in human beings cause pain in other vertebrate species, although more needs to be known about the perception of pain in animals.

VII. Procedures with animals that may cause more than momentary or minimal pain or distress should be performed with appropriate sedation, analgesia, or anesthesia in accordance with accepted veterinary practice. Surgical or other painful procedures should not be performed on unanesthetized animals paralysed by chemical agents.

VIII. Where waivers are required in relation to the provisions of article VII, the decisions should not rest solely with the investigators directly concerned but should be made, with due regard to the provisions of articles IV, V, and VI, by a suitably constituted review body. Such waivers should not be made solely for the purposes of teaching or demonstration.

IX. At the end of, or, when appropriate, during an experiment, animals that would otherwise suffer severe or chronic pain, distress, discomfort, or disablement that cannot be relieved should be painlessly killed.

X. The best possible living conditions should be maintained for animals kept for biomedical purposes. Normally the care of animals should be under the supervision of veterinarians having experience in laboratory animal science. In any case, veterinary care should be available as required.

XI. It is the responsibility of the director of an institute or department using animals to ensure that investigators and personnel have appropriate qualifications or experience for conducting procedures on animals. Adequate opportunities shall be provided for in-service training, including the proper and humane concern for the animals under their care.

2. SPECIAL PROVISIONS

Where they are quantifiable, norms for the following provisions should be established by a national authority, national advisory council, or other competent body.

2.1 ACQUISITION Specialized breeding establishments are the best source of the most commonly used experimental animals. Nonspecifically bred animals may be used only if they meet the research requirements, particularly for health and quality, and their acquisition is not in contradiction with national legislation and conservation policies.

2.2 TRANSPORTATION Where there are no regulations or statutory requirements governing the transport of animals, it is the duty of the director of an institute or department using animals to emphasize to the supplier and the carrier that the animals should be transported under humane and hygienic conditions.

2.3 HOUSING Animal housing should be such as to ensure that the general health of the animals is safeguarded and that undue stress is avoided. Special attention should be given to the space allocation for each animal, according to species, and adequate standards of hygiene should be maintained as well as protection against predators, vermin, and other pests. Facilities for quarantine and isolation should be provided. Entry should normally be restricted to authorized persons.

2.4 ENVIRONMENTAL CONDITIONS Environmental needs such as temperature, humidity, ventilation, lighting, and social interaction should be consistent with the needs of the species concerned. Noise and odour levels should be minimal. Proper facilities should be provided for the disposal of animals and animal waste.

2.5 NUTRITION Animals should receive a supply of foodstuffs appropriate to their requirements and of a quality and quantity adequate to preserve their health, and they should have free access to potable water, unless the object of the experiment is to study the effects of variations of these nutritional requirements.

2.6 VETERINARY CARE Veterinary care, including a program of health surveillance and disease prevention, should be available to breeding establishments and to institutions or departments using animals for biomedical purposes. Sick or injured animals should, according to circumstances, either receive appropriate veterinary care or be painlessly killed.

2.7 RECORDS Records should be kept of all experiments with animals and should be available for inspection. Information should be included regarding the various procedures which were carried out and the results of post mortem examinations if conducted.

3. MONITORING OF THE CARE AND USE OF ANIMALS FOR EXPERIMENTATION

3.1 Wherever animals are used for biomedical purposes, their care and use should be subject to the general principles and criteria set out above as well as to existing national policies. The observance of such principles and criteria should be encouraged by procedures for independent monitoring.

3.2 Principles and criteria and monitoring procedures should have as their objectives the avoidance of excessive or inappropriate use of experimental animals and encourage appropriate care and use before, during, or after experimentation. They may be established by: specific legislation laying down standards and providing for enforcement by an official inspectorate; by more general legislation requiring biomedical research institutions to provide for peer review in accordance with defined principles and criteria, sometimes with informed lay participation; or by voluntary self-regulation by the biomedical community. There are many possible variants of monitoring systems, according to the stress laid upon legislation on the one hand, and voluntary self-regulation on the other.

4. METHODS NOT INVOLVING ANIMALS: "ALTERNATIVES"

4.1 There remain many areas in biomedical research which, at least for the foreseeable future, will require animal experimentation. An intact live animal is more than the sum of the responses of isolated cells, tissues or organs; there are complex interactions in the whole animal that cannot be reproduced by biological or nonbiological "alternative" methods. The term "alternative" has come to be used by some to refer to a replacement of the use of living animals by other procedures, as well as methods which lead to a reduction in the numbers of animals required or to the refinement of experimental procedures.

4.2 The experimental procedures that are considered to be "alternatives" include non-biological and biological methods. The nonbiological methods include mathematical modeling of structure-activity relationships based on the physico-chemical properties of drugs and other chemicals, and computer modeling of other biological processes. The biological methods include the use of micro-organisms, in vitro preparations (subcellular fractions, short-term cellular systems, whole organ perfusion, and cell and organ culture) and under some circumstances, invertebrates and vertebrate embryos. In addition to experimental procedures, retrospective and prospective epidemiological investigations on human and animal populations represent other approaches of major importance.

4.3 The adoption of "alternative" approaches is viewed as being complementary to the use of intact animals and their development and use should be actively encouraged for both scientific and humane reasons.

NOTES

CHAPTER 1

1. For an essay comparing the Arab Spring and Occupy Wall Street movements and relating them to civil-society revolutions of 1989 and 1968, see Kennedy 2011.

2. Sometimes I use "human embryonic stem cell (hESC) research" and "human pluripotent stem cell research" synonymously. Sometimes, depending on the context, I use one and not the other because I am talking about a time before or after the proof-of-principle papers on induced pluripotency had been published. (See chapters 2 and 3.) Sometimes I use one term and not the other because I am talking specifically about induced pluripotent stem (iPS) cells derived from sources other than embryos, or because I am talking specifically about embryonic stem cells derived from embryos. "Pluripotent stem cells" (abbreviated PSCs) is the collective term.

3. For an example of the ambivalent or multi-layered way in which the United Nations and its agencies relate to majoritarianism, deliberation, consensus, the protection of minority rights, and engagement with civil society and the private sector, in its formal structure and its decision-making processes, see "Section One: Decision-Making at the United Nations: How It Works" in United Nations Non-Governmental Liaison Service 2003.

4. Sheila Jasanoff (2005) discusses this latency of dissent and controversy, as well as the national characteristics of the forms it tends to take in different places.

5. Elsewhere I elaborate on my use of the word "ethics" and its relation to other fields of scholarship and practice.

6. See Thompson 2007a.

7. See Yusa et al. 2011.

8. An editorial in the October 6, 2011 issue of the journal *Nature* ("High-interest clones") summarized some aspects of the ongoing debate about the need for gametes and embryos, varying views on egg-donation protocol, and the similarities

and differences between somatic cell nuclear transfer as claimed by Hwang Woo-Suk in 2004 and 2005 (Hwang et al. 2004 and 2005; an archive of the investigation and retraction of these two papers is available at http://www.sciencemag.org), and by the New York Stem Cell Foundation Laboratory in 2011 (Noggle et al. 2011).

9. See Thompson 2012.

10. Here I am building on my own work on "the biomedical mode of (re)production" (Thompson 2005) and on egg donation (e.g., Thompson 2007b, 2009), and on the exceptional work of many others on the stratified moral and monetary economies for the transfer of bodily parts and labor in biomedicine (e.g., Browner and Sargent 2011; Cohen 2002 and 2005; Cooper 2008; Ginsberg and Rapp 1995; Petryna 2009; Roberts 1999 and 2003; Scheper-Hughes 2005, 2006; Scheper-Hughes and Wacquant 2002; Sunder Rajan 2006; Teman 2010; Vora 2009; Wailoo 1997; Waldby and Mitchell 2006).

11. See the work on "vital" ethics and politics by Sarah Franklin (now at Cambridge University) and Nikolas Rose (now at King's College London), both then of the BIOS Center at the London School of Economics—e.g., Franklin 2007 and Rose 2006. See also Frank Hendriks' (2010) notion of "vital democracy." In a U.S. context, thinking about "vital" politics, regardless of one's approach to the politics of the life sciences, brings to mind Arthur Schlesinger's (1949) Cold War defense of "the vital center." While I share with Schlesinger the sense that science and technology are crucial to the discontent as well as the promise of the modern age, the physics of the Cold War period and the life sciences of the early twenty-first century make for very different conditions. Schlesinger's use of nationalized gendered epithets to defend freedom from extremists of the right and left (bad feminized men and good American masculinity; see Penner 2010, pp. 67–97), and his desire for centrist stability, underline the connections in the "atomic-age" U.S. between liberalism and the hierarchical gender system of the postwar nuclear family that was to be challenged by the civil-rights, feminism, and gay-liberation movements of the 1960s and the 1970s. In the current age of *in vitro* embryos and finance oligarchy (Johnson 2009), by contrast, the family is no longer only nuclear, and the center is no longer a safe haven for political stability and freedom. Hybrid and plural democratic forms are today as essential to ethics of the life sciences as they are to a living, vibrant democracy in a society whose citizens are increasingly caught up in the life sciences.

12. Thomson et al. 1998.

13. Yu et al. 2007; Takahashi et al. 2007.

14. On ethical issues around the emergence of stem cell research, see Holland, Lebacqz, and Zoloth 2001. On the period around California's Proposition 71, see Scott 2006 and Benjamin 2013. On the emergence of pluripotent stem cell politics more globally, see Gottweis, Salter, and Waldby 2009.

15. Ted Peters (a theologian, a member of the Geron Corporation's Ethics Advisory Board, and a member of the Scientific and Medical Standards Accountability Working Group at the California Institute for Regenerative Medicine) characterizes three moral frameworks in the California stem cell debate: the embryo-protection framework, the nature-protection framework, and the medical-benefits frameworks (Peters 2007). His typology provides ethical reasons not only for adhering to each of the frameworks, but also for being unable to engage the other frameworks from within any single framework.

16. Among the works that theorize ongoing redefinitions of life, death, reproduction, and health in biomedicine are Clarke et al. 2010, Epstein 2007, Franklin and Lock 2003, Haraway 1990, Kaufman 2006, Lock 2001, Lock and Nguyen 2010, Petryna 2002, and Rapp 2000. See also the works cited in note 10 above.

17. On the effect on one's research of "mutual entanglements," see Callon and Rabeharisoa 2004. On a turn to "matters of concern" in science and technology studies, see Latour 2004.

18. See, e.g., Galison 2004a, a work that illustrates both the contingency of the different paths of science and the essential materiality connecting what has come before, the research of today, and imagined futures for a field.

19. On the importance of political legitimacy to science, and vice versa, see Jasanoff 2004 and Jasanoff 2005. Equally important is Jasanoff's empirical work on the differences between countries in the forms that the relationship between political legitimacy and science can take. Scientific futures cannot be created *de novo*; they must make sense in terms of the repertoire available in a given place.

20. This difference between reproductive technologies and human embryonic stem cell research is evident in social-scientific scholarship about the fields, too. Reproductive technologies are familiar terrain for feminist theory; stem cell research is less so.

21. As Sarah Franklin has shown (2005, 2006), human pluripotent stem cell research has developed in greater proximity to the fertility industry and *in vitro* fertilization in the U.K., because of the U.K.'s long history of publicly regulating research and therapy involving human gametes and embryos in the context of fertility medicine.

22. In some ways my research trajectory was the opposite of Charles Bosk's. Bosk (2008) followed professional bioethicists *in situ* and wrote about what that meant to an ethnographer who had plenty of his own insights into questions that feature in bioethics in codified if open-ended ways. Human embryonic stem cell research was not yet clinical, and the bioethics was still in some ways up for grabs.

23. See the essays in Moreno and Berger 2010; see also Moreno 2011.

24. See Jasanoff 2005 for her "use [of] the term 'ethnography' analogically" to find in such things as bureaucratic and institutional forms a society's "self-perpetuating normative commitments."

25. On triage protocol in U.S. emergency rooms, see Gilboy et al. 2005. Different countries and institutions have guidelines for triage in conflict zones, during peacekeeping operations, during famines, during epidemics, and for emergency-room and hospital treatment, and it is generally agreed that those who could live but only with immediate treatment take precedence over those who will die anyway or those who can afford to wait; in general, the labeling systems reflect this. When resources are too limited to treat everyone who could be saved, criteria such as the likelihood of battle-readiness after treatment can be called upon; other criteria, such as projected remaining life span, likely quality of remaining life, or likely cost or perceived value to society, are disputed.

26. On the relation between triage and governance in biomedical geopolitics, see Nguyen 2010. Visvanathan (1997) uses the matched concepts of vivisection and triage to explore "Western" science's legacy of experimenting upon and stratifying people, placing vivisection and triage at the heart of the modern nation, of the colonial relation, and of modern science, and of the integral and co-productive relations among all three.

27. On the ways in which research has been integral to the colonial project and the problems with representing others in voice or otherwise, see Smith 1999. According to Smith, "the word itself, 'research' . . . stirs up silence" (ibid., p. 1).

28. See Juana Rodriguez's (2003) exploration of the always evolving, multivocal, and contested categories of identity politics, and of subjectivity itself.

29. The comments quoted in the paragraphs that follow are drawn from notes, and thus often only represent my best recollection; they are, however, often ver-batim or close to it. I have given no uniquely identifying information, and each comment was made to me or to a group that included me in an open venue (a classroom, group office hours, a formal session or group conversation at a meeting or conference or public lecture, or a written comment summarizing publicly shared views). I have not reproduced any comments made in a venue that might have been supposed to be private (one-on-one office hours, a friend's or a col-league's private written or verbal communication, a comment made in confidence at a public venue).

30. On "multiple biomedical ontologies," see Mol 2002. On the different "epis-temic cultures" of different scientific disciplines, see Knorr Cetina 1999. On the differences within a single discipline in how different countries define and pursue the same field, see Fourcade 2009. On how cross-fertilization between scientific fields can occur, see Galison 1997.

31. On ethnography beyond the single site, on "partial connections," comparison, complexity, the problems of scale, and the productivity of reading across domains, see Strathern 2005. Clarke and Oleson (1998) draw on Donna Haraway's (1990b) "situated knowledges" to demand a framework that they call "Revising, Diffract-ing, and Acting," illustrating that a feminist understanding of the (multi-) situated-

ness of science helps rather than hinders analysis and advocacy across domains and recalling the feminist motivations for developing situated knowledges in the first place: showing the embodiment of actors, the partiality of perspective in knowledge claims, and who should take responsibility for and who gains and who loses in hierarchies of power that overflow the local. On "the field site as a network" and the network as "a strategy for locating ethnographic work," see Burrell 2009.

32. Alarcon, Kaplan, and Moallem (1999, p. 4) have argued that expectations of how one should proceed methodologically and theoretically are themselves historically contingent legacies of modernity which it is imperative yet difficult to destabilize precisely because they carry with them the stamp of epistemic authority. They "argue for closer attention to . . . transnational conditions of knowledge production and consumption" if we are not to get stuck "within the discursive cosmos of colonial power relations, helpless to recognize the complex and nuanced manifestations of transnational circulations of peoples, goods, and information in the present moment."

33. Recent queer theory and critical race theory on archives of feeling and affect theory is suggestive for thinking about the affect-laden stem cell research archive. See, e.g., Cvetkovich 2003 and Eng 2010.

34. Transcript triage, in that it reads against the logic of the mainstream archive in question, bears some similarity to the idea of the "hidden transcript." For James Scott (1992, p. 4), "transcript is used almost in its juridical sense (process verbal) of a complete record of what was said" while the hidden transcript is used "to characterize discourse that takes place 'offstage,' beyond direct observation by powerholders." Evelyn Nakano Glenn's intersectional re-working of the idea of hidden transcript is also very helpful here (2004). For feminist writings on the colonial archive, see Arondekar 2009 and Stoler 2009.

35. Giorgio Agamben (1998) grounds biopolitics in the unspeakable atrocity of the Holocaust and considers what can and cannot be witnessed or archived at the limits of humanity. The stem cell archive is haunted in its biopolitics by the suffering and ticking clock of patients; the medically relevant consequences of who has access to treatment or which research priorities are pursued casts only a faint shadow on the field's archive.

36. Lynch (1999) builds on Derrida's (1996) work on the archive, from its ancient etymology up through email, and demonstrates the range of archival formats of our digital and visual media age and their different logics of and contests over their collection, writing, deleting, and interpreting. On digitality, also see Trinh 2005.

37. Laura Perez (2007) has argued for and undertaken the political task of having a book function as an archive when the archive in question is ephemeral or counter-hegemonic or in other ways not easily seen.

CHAPTER 2

1. On the rise of the "bioethical enterprise" in U.S. medical history, see Rosenberg 1999. On the role of Mary Warnock in the rise of bioethics, and its epithet "industry" in the U.K., see D. Wilson 2011. On the growth in kind and scale of "bureaucracies of virtue," see Riles and Jacob 2007.

2. On the framing and overflowing of markets, and on what counts as an "externality" (and instability in that), see Callon 1998a. On "the biomedical mode of (re)production," and on what it takes to extend capitalist-like markets to transnational capitalized reproductive, recombinant, and regenerative bodies and body parts, see chapter 8 of Thompson 2005.

3. On biomedicine's rise in conjunction with the chemical industry, see Lesch 2007. On the emergence of a major post-1970s and 1980s recombinant DNA company, Genentech, and its struggles to make a genetic engineering market, see Smith-Hughes 2011.

4. On the role of gender, race, immigration, and transnational capital in limits to the commodification of care work, see Glenn 2010. On the role of market logic in keeping emotion and bodies outside markets, see Zelizer 2010 and Hochschild 2003. On the rise of markets enabled in part by the rhetoric of resistance to the market, see Spar 2006.

5. At times, ethics not only makes it permissible to do work in sciences that have ethics; it can also be the very technology of creating and maintaining wage gradients and labor pools. On ethical variability, and on ethics as a "workable document," see Petryna 2009.

6. Although contributing to work in the ethics of science and biomedicine, this book is also *about* the phenomenon and proliferation of something called "ethics" and/or "bioethics" in the governance of the biological science. The latter is an important recent topic of empirical and theoretical investigation in the social sciences. See Nowotny et al. 2001, Haimes 2002, and Fox et al. 2008.

7. The point about "post-humanism" is that once we can re-engineer, informatize (turn into information), and protheticize (add enhancements or surrogates to) the body and the mind, the essence (whatever that might ever have been; the idea of an essential difference has long been highly contested at the margins of human/machine and human/animal, and regarding modern classificatory systems such as gender, race, and sexuality, for example) of each human or of humanity in general disappears. See, e.g., Baillie and Casey 2004; Malil and Cheng 2011. It is crucial to note, however, that these same converging sciences that have ethics are also re-inscribing narratives of identity based on modernist, humanist categories. See, e.g., TallBear 2012 and Bolnick et al. 2007.

8. Not all ethical controversy around science that has a ready-made constituency is so closely aligned with party politics as embryo destruction in the U.S.; for

example, in the U.K. animal rights demands public attention much as abortion does in the U.S., but is not as closely aligned with party politics.

9. For a review and an analysis of this literature, see chapter 1 of Thompson 2005.

10. Several scholars have called for a new social contract for science in society during the period of research covered in this book. For an earlier discussion (U.K.-based; the calls for a new social contract for science in this period came first from Europe; see Nowotny et al. 2001) of the need for a new social contract because of changing links between university and industrial science, see, e.g., Gibbons 1999, and for a later (U.S.-based) call for a new social contract to meet the special challenge posed by sciences that rely on human tissue and bio-information, see, e.g., Meslin and Cho 2010.

11. For the counter-argument to this position, see chapter 4 of Thompson 2005.

12. Emphasis added. The ELSI acronym in the U.S. dates to 1989, in the early stages of the Human Genome Project.

13. Source: National Human Genome Research Institute, "What is ELSI?" (available at http://www.genome.gov).

14. During the fieldwork for this book, UC Berkeley became embroiled in a controversy over whether or not to accept funding from the petrochemical firm BP. As an indication of how far the life sciences have come to be associated with private profit and how fiercely university administrators and certain researchers have normalized and are promoting that state of affairs, the majority of scientists on campus (including many of the scientists in our Berkeley Stem Cell Center, who discussed the case with me) came down in favor of accepting the money, while humanists and social scientists were left to raise questions about the public university and the threats of such a deal to its character. By staging the discussion as a partisan vote, vital questions about research freedom were shut down. On "upstream ethics" in nanotechnology, and for an analysis of the stakes of keeping ethics "downstream," see Khushf 2007.

15. Since the beginning of the Human Genome Project's ELSI working group, and the proliferation of ELSI programs in other areas, a persistent theme in U.S.-based ELSI researchers' and others' accounts has been that ELSI research was sidelined or ignored. On the various ways in which ELSI in conjunction with the HGP was funded but kept at arm's length, see Allen 1997. On the ways certain emerging pathways in synthetic biology have constructed the "human practices" component of research as an externality to the research itself, see Rabinow and Bennett 2009. Scientists struggled for cultural legitimacy (see, e.g., Snow 1959) in other times and other places, and may do so again.

16. On the politics of "spare embryos," see chapter 8 of Thompson 2005. On the complex moral landscapes of the spare embryo that emerged after the legalization of hESC in Denmark, see Svendsen and Koch 2008. For an equally telling look at the ambiguities and treatment stage dependent aspects of labeling an embryo

"spare" that led to the use of frozen rather than fresh donor embryos in the context of U.K. efforts to ease the IVF/hESC interface, see Ehrich, Williams, and Farsides 2010.

17. SCNT was the procedure used to create Dolly the sheep in 1997, and which Hwang Woo Suk (see next chapter) claimed to have carried out for humans; it uses an egg which is subsequently enucleated and into which the DNA of a somatic cell from a patient would be placed. Human SCNT, if technically feasible, would enable the creation of patient-specific stem cells that would not trigger an immune response in the patient. Procuring the eggs from women for SCNT provoked a range of additional ethical debates, however, such as whether or not donors should be paid for donating the eggs needed for the procedure and the possibility that therapeutic cloning might lead to reproductive cloning: cloning for the purpose of creating entire individuals as opposed to producing tissue.

18. Senate Bill No. 1260, Ortiz, 2006 Law, available at http://www.cdph.ca.gov.

19. Egg donation is dealt with at greater length in chapter 3.

20. Official White House transcript of President Bush's remarks on May 24, 2005. Text available at http://www.thefreelibrary.com.

21. On experiences deriving human embryonic stem cell lines from "non-viable" fertility treatment embryos at Columbia University, see Garilov et al. 2011. Empirical work in clinics has shown how variable in practice different clinicians' judgments of viability and non-viability are, and how patients' stage and outcome of treatment and other characteristics effect this designation.

22. See Klimanskaya et al. 2006.

23. See Thompson 2007b.

24. See Hurlbut 2005.

25. Wishing to signal the compatibility of Catholic respect for life with both adult stem cell sciences and their commercialization, the Pontifical Council for Culture at the Vatican announced the furthering of its partnership with NeoStem, Inc., a U.S. company that works in the U.S. and China with adult stem cells, in June 2011. The Vatican hosted an international conference on Adult Stem Cells: Science and the Future of Man and Culture on November 9–11, 2011.

26. See "World Stem Cell Policies," available at http://www.hinxtongroup.org.

27. See "Freezing Embryos" in "Second Part: New Problems Concerning Procreation" (Congregation for the Doctrine of the Faith 2008).

28. See chapter 4 below.

29. See appendixes C and D.

30. The intent behind the 2011 amendments was confirmed by Geoff Lomax of CIRM in his compliance visit to UC Berkeley's SCRO committee, on which I served, on May 19, 2010.

31. The California Department of Public Health followed suit in its December 2011 guidelines for doing stem cell research in California, binding on all pluripotent stem cell research in the state except that entirely funded with CIRM funds. Their revisions formalized the exception of almost all kinds of iPSC research from "covered research" and hence exempted it from SCRO review. Version December 2011, California Department of Public Health Guidelines for Human Stem Cell Research, as recommended by the Human Stem Cell Research Advisory Committee, pursuant to Health and Safety Code 125118, available at http://www.cdph.ca.gov.

32. By using the word "rhetoric," I do not mean to imply that supporters were only pretending to care about cures; supporters care passionately about cures. I am interested, here, in how support for the research was expressed to overcome pro-life objections and to secure a massive state-funded initiative dedicated to this particular field of research.

33. The text of Proposition 71 can be found at http://www.cirm.ca.gov/sites/default/files/files/about_cirm/prop71.pdf. Proposition 71 was approved by 59.05% of the vote, the turnout exceeding 76% of California's more than 16.5 million registered voters, in November 2004.

34. For the U.S. hESC context of biomedicine knowledge economy market development in the face of an "ever-present" possibility of market failure, see Salter 2010.

35. See Åhrlund-Richter et al. 2009.

36. Thomson et al. 1998.

37. See chapter 5 below.

38. On hESC "disease in a dish" research from around the period of Proposition 71, see Pickering et al. 2005. On disease-specific iPS cells, see Park et al. 2008.

39. See Scott 2008.

40. See Engelberg, Kesselheim, and Avorn 2009.

41. See Lengerke and Daley 2010.

42. On November 14, 2011, when the Geron Corporation called a sudden halt to a hESC-based clinical trial (for which it had already enrolled five patients, and toward the funding of which CIRM had controversially awarded it a $25 million loan) and abandoned hESC work in favor of oncology work, it cited financial reasons. Reporters pointed out that the financial reasons, uncertainties about the science, the unrealistic hope/hype way in which the California stem cell initiative was structured from the start (and to which the recently departed former CEO of Geron, Thomas Okarma, had been central), and the future of CIRM itself were all caught up in the fallout. See Hiltzik 2011 for these interrelated strands. For scans of the Geron-CIRM loan documents and documentation of the episode, see Jensen 2011.

43. At a 2006 meeting hosted by the California Institute for Regenerative Medicine for the for-profit sector that I attended, speakers confirmed that biotech venture capitalists don't like investing in science when the ethical, regulatory, or legal environment is risky, or when the business model appropriate to the kind of science is unclear.

44. See Hayden 2008 for an analysis of the dual strands of patient advocacy and private philanthropy at the heart of CIRM and its science, including the role of Robert Klein in moving between local and national politics and private and public funding.

45. Under California's Proposition 13, passed in 1978, businesses and long-term residents who have not moved and their children pay a property tax based on the value of their homes in 1975 (before the housing boom). Newer homeowners must pay a rate based on what they paid for their houses whenever they bought them. California's proposition process is a form of direct democracy or populism that at its best shakes up politics as usual in the state, but a measure passed as a proposition is hard to repeal (that requires a 70% majority) and can have serious unintended consequences for the common good down the line.

46. See, e.g., Epstein 1996 and Ganchoff 2008.

47. On biosociality, see Rabinow 1992 and Rose 2006.

48. This struck me repeatedly during my research.

49. Available at http://www.lao.ca.gov/ballot/2004/71_11_2004.htm.

50. The Center for Genetics and Society and the Foundation for Taxpayer and Consumer Rights (subsequently renamed the Consumer Watchdog, California) Stem Cell Oversight and Accountability Project pointed out the potential for conflicts of interest and the lack of benefits to the public as early as 2005; see Center for Genetics and Society 2005 and Flanagan 2005. David Jensen, the author of a blog called The Stem Cell Report (http://californiastemcellreport.blogspot.com/), has documented these issues in his comprehensive California stem cell initiative archive since 2005.

51. See Little Hoover Commission 2009.

52. On tensions between reliable knowledge production and the "desperation" of patients, see Lowry 2000. On "the monopoly of desperation," see chapter 7 of Thompson 2005. On the construction of the desperateness of patients, see chapter 5 of Franklin 1997.

53. For what is more than "an advanced introduction" to biopolitics, see Lemke 2011.

54. For a feminist, ecological genealogy of the concept, see Shiva and Moser 1995.

55. Foucault 1998.

56. Casper and Moore 2009.

57. On the concentration camp as the scenography of biopolitical abjection, see Agamben 1998. On the necropolitical, see Mbembe 2003. On the question of who can and who cannot be mourned in the post-9/11 era, see Butler 2006.

58. Clough and Willse 2011.

59. On "lively capital," see Rajan 2012.

60. Here, to capture the aspect of biomedicalization that casts the living as always already previvors and survivors of personal biomedical identities, I draw on Sarah Jain's (2006) concept of "living in prognosis."

61. The intertwining of these competing bioscapes is clear; regardless of which position is uppermost at a given time or place during this period in the U.S., the other position makes its presence felt. President George W. Bush's federal policy spurred state initiatives to fund or permit or more strongly prohibit hESC, not just in California but also in several other states—for example, New Jersey, Connecticut, Illinois, New York, Maryland, Iowa, and Massachusetts passed laws to fund embryonic stem cell research; Indiana and Virginia created adult stem cell initiatives instead. See Embryonic and Fetal Research Laws, National Conference of State Legislature, available at http://www.ncsl.org. Likewise, President Obama's 2009 easing of restrictions for NIH funding of research sparked a long-running lawsuit (*Sherley v. Sebelius*), just as the passage of Proposition 71 did in California.

62. Recent feminist and queer theory has elucidated ways of being outside the archive that are germane here. Jack Halberstam (2011) draws on "silly archives" and "low theory" to inhabit the landscapes that refuse success (bio-innovation is a primary site of "success" mongering). Sara Ahmed's (2010) "unhappy archives" trace the biopolitical and necropolitical cross-cutting terrains of "feminist killjoys," "unhappy queers," and "melancholy migrants," illegible from the realm of hegemonic "happiness" as the individual telos. Trinh Minh-ha (2005) illustrates many ways and registers and senses in which to count and be counted, and Mel Chen (2012) fruitfully deploys the linguistic notion of animacy to trace the materialities of matter and mattering.

63. On the politics of race, gender, and disability in California's stem cell initiative, and the efforts of various groups and actors to bring issues of social justice and health disparities into the conversation, see Benjamin 2013.

64. On the various forms of potential economic gains, see Tolley 2008.

65. CIRM was operating on loans while waiting for the bonds to be freed from litigation; on July 20, 2006, Governor Schwarzenegger announced that the state of California would buy $150 million of CIRM's bonds, and the financing board authorized $200 million to be sold, leaving $50 million to be bought privately, in advance of the bonds being available for sale to the general public because of the pending litigation.

66. UC Berkeley's Science, Technology, and Society Center (STSC), of which I was then director, and the Berkeley Stem Cell Center co-sponsored the Toward

Fair Cures conference with the Greenlining Institute (STSC and Greenlining conceptualized and organized and funded the event) and the Children's Hospital Research Institute, CHORI (which hosted the event). It took place at CHORI on November 15, 2006. Dr. Hall sent a video message for the occasion. CIRM's Scientific Review Officer, Gil Sambrano, also participated.

67. For example, the following was my public comment on one such skirmish, which concerned a facilities grant:

PUBLIC COMMENT, SUBMITTED BY Charis M Thompson, UC Berkeley
January 16, 2008
Dear Members of the ICOC,
As you consider which applications for CIRM funding for facilities grants to approve today and tomorrow, I urge you to recall that CIRM came about as a result of a public mandate for research and cures. Unlike funding for science that comes through the private sector, through the military, or even through the tax base mediated through such federal agencies as NSF and NIH, CIRM funding came about through a popular bond initiative, Proposition 71, focused on providing research and cures for the people of California. It could be argued that CIRM granting should thus demonstrate a particular sensitivity to serving the people of California. This means that distributive factors should be taken into account that might mean that the final group of proposals funded under any RFA does not exactly mirror the recommendations of scientific or other proposal review groups who were not asked to take this particular public mandate of CIRM's into account. It also means that efforts must be made to mitigate an inherent unfairness in this call for proposals, namely the stated intention to favor in the next round those institutions that can raise matching funds for facilities. This invites bias toward proposals coming from those institutions that are already the richest and have the richest donor base.

As a concerned member of the public, and as an academic in the field of bioethics, I urge you to consider:
a) Striving to fund a balance of facilities that will widely serve the people of California
b) Setting out in at least some instances to fund facilities for which there is not likely already to be a large and willing donor population
Thank you for your attention.
Charis M Thompson

68. "Americans for Cures" was the name given to the 501(c) 3 (charitable status) organization that was formed the day after the passage of Proposition 71; during the campaign, it was the "Yes on 71" group the Alliance for Stem Cell Research.

69. SB 1565, a bipartisan California Senate Bill, was introduced by Senators Kuehl and Runner on February 22, 2008.

70. Available at http://www.americansforcures.org.

71. California Senate Bill SB 1064, introduced by Elaine Alquist (D-Santa Clara), was approved by the governor and filed with the secretary of state on September 30, 2010, having been endorsed unanimously by CIRM's governing board, the Independent Citizen's Oversight Committee (ICOC). It improved public oversight of CIRM, ensured that some monies from any future earnings would flow back to the state, and ensured that standards of affordability of resultant treatments would be met.

72. The text of SB1064 is available at http://www.leginfo.ca.gov.

73. See, e.g., Holloway 2011, which calls for a cultural bioethics that pays attention in the very framing to the social and historical factors that render some populations more vulnerable than others in medical contexts. See also http://womensbioethics.blogspot.com/.

74. On the questionable ethics of stopping the trial for business reasons, see Bayliss 2011.

75. The British disability-rights movement differentiated a medical model of disability from the social model in the 1970s. A medical model places disability in and on the individual, and emphasizes medical solutions to "fix" impairment; the social model separates impairment from disability itself, conceptualizing disability as a social power relation between individuals living with impairment and the society in which they are discriminated against. The social model seeks the removal of social and physical barriers to full participation (pointing out that stairs create the disability of being unable to go above the ground floor for someone in a wheelchair, but a ramp or an elevator does not). See, e.g., Oliver 1990. Over the years, disability-studies scholars have complicated the relations between the medical and social models of disability, particularly as regards taking seriously the role of medicine for many living with disabilities, even while insisting on the continued political importance of the social model of disability. For a review of work on the distinction, see Shakespeare and Watson 2002.

76. The disability-justice rallying call "Nothing about us without us" is apt here. It is harder to apply it to the politics of patient advocates as spokespeople for public support of stem cell research, however; rhetorically the latter might more accurately be rendered as "Nothing for us without you, and nothing for you without us," which has a slightly different valence.

77. In an important change, the World Health Organization International Classification of Functioning, Disability, and Health (ICF) of 2001, endorsed at the 54th World Health Assembly, replaced the older International Classification of Impairments, Disabilities, and Handicaps (ICDH); unlike its predecessor, it places disability, functioning, and health on a spectrum and in social and environmental context.

78. On stem cell ethics, I have been especially influenced by the work of Patricia Berne, who gave a powerful lecture to one of my classes on the social and medical models of disability and the imperative of intersectional (disability, class, race, sexuality) analysis, and Adrienne Asch and Marsha Saxton, who have led the way in navigating reproductive choice, access to medical care, and eugenics in the newer biomedical technologies. See, e.g., Berne 2008; Asch 2000; Saxton 2000. Working on stem cell topics with students living with a range of disabilities (including blindness and cerebral palsy), with conditions such as diabetes, and with students and colleagues caring for individuals with from-birth cognitive disabilities or end-of-life dementia also played an important in shaping my thought.

79. On the rise of these concerns and terms of debate, see Thompson 2007b.

80. See Thompson 2012.

81. For example, through PromoCell, I can purchase 500,000 cryopreserved normal human dermal fibroblast cells from "juvenile foreskin" for 650 euros, or $436, without knowing which little boys donated these cells, how consent on their behalf was handled, how much it is reasonable to charge, or who should see those profits. As long as the consenting and business and lab certification and credentialing practices are within the law and accepted practice, nothing need be specified about them. Burying them under "business as usual" makes it harder to bring them to the light of day for possible reform.

82. Brynne, Nguyen, and Reijo Pera 2009.

83. See other papers on iPSC from the same period—e.g., Huangfu et al. 2008, which gives the following on the provenance and procurement of human cell lines for the iPSC study: "Human BJ (ATCC) and NHDF (Lonza Biosciences) cells were maintained in fibroblast medium." "Human BJ" refers to a line of human foreskin fibroblasts that is longer lived than normal fibroblasts. "ATCC" refers to a major biorepository in Virginia that has operated as a private non-profit source of human cell lines since 1925 and that ships cell lines around the world. "NHDF" refers to human dermal fibroblasts. Lonza Biosciences is a major Swiss biopharmaceuticals supplier. The company and the kind of cells being used stands in for all other information about the provenance, procurement, or disposition of the human cells being bought for use in research.

84. For example, Woltjen et al. (2009) simply refers to the provenance of human embryonic fibroblast cells (which are "adult" cells in the language of the stem cell debate because they are differentiated) as follows: "HEFs were derived from 12-week abortuses."

85. Available at http://www.atcc.org.

86. See the summary in Shea 2004. This passage is not atypical in prevaricating as to whether it is the civilian (public health) or military (biological weapons) potential that makes the research dual-use, or the fact that biological weapons research can be both defensive and offensive that makes it dual-use.

87. Galison 1994.

88. On the "classified theory of knowledge," see Galison 2004a.

89. On the co-construction of new technologies of war and peoplehood, see Kaplan 2006. On how war targets and simultaneously constructs gendered, ethnoracialized, nationalized, religious categories of peoplehood from which we may know who is the enemy and who is being protected, see Zarkov 2007. On the concept of "modern peoplehood," see Lie 2004.

90. Till and McCullough 1961.

91. See Furcht and Hoffman 2008, which includes a chapter on the potential role of stem cell research in biodefense. Also see Hall 2008, in which the founding director of CIRM, Zach Hall, discusses the importance of adding this dimension to the debate.

92. Most public funding for biodefense in the U.S. comes through the various agencies of the Department of Defense, with some funds coming through the National Institutes of Health. Because of secrecy, public engagement with this research is more likely to be downstream, through technology diffusion, than through public debate about appropriations or the use of funds for these purposes.

93. Executive Office of the President of the United States 2007.

94. Ibid.

95. For a sense of the history of links between biomedical research and the military in the Cold War, see Welsome 2000. On the links between national defense and brain science, see Moreno 2006.

96. I have heard civil-society groups such as the Center for Genetics and Society described as anti-science, despite the fact that they are explicitly pro not only science in general but human pluripotent stem cell research in particular.

97. National Bioethics Advisory Commission 1999; Thomson et al. 1998.

98. National Bioethics Advisory Commission 1999, p. 2.

99. For an overview of the concept of human dignity in relation to U.S. bioethics in the PCB, see President's Council on Bioethics 2008. For the PCB's work on alternative (non-embryonic) sources of human multipotent stem cells, see President's Council on Bieothics 2005.

100. Although some might argue for or against particular deontological or consequentialist or communitarian or practical or situational ethical frameworks as *the right way* to resolve the stem cell debate, that is not my aim here. All these ethical positions were evident in different settings among different actors, as regards different issues, around stem cell research in the U.S. in this period. It is the ethical choreography, or the undoing and re-attaching of bundles of human and nonhuman actors making up interests, upon which I am focusing in this book.

101. For a call for praise from across the aisles for the PCB's work, see Elliott 2004, in which the author hailed the depth and thoughtfulness of the Council's report *Beyond Therapy: Biotechnology and the Pursuit of Happiness* (President's Council on Bioethics 2003). Partisan interpretations of the dismissals of William May and Elizabeth Blackburn from the council for allegedly clashing with the PCB's chair, Leon Kass, were more common in the news media.

102. Source: Carter 2009 (first published at http://www.firstthings.com). This article was reproduced in the Catholic blog Pro-Ecclesia, Pro Familia, Pro Civitate (http://proecclesia.blogspot.com) as one of its Obama Culture of Death Updates.

103. The announcement by the Federation of American Scientists, "New Bioethics Council to Advise the White House" (made at http://www.fas.org), said the following: "The Council, appointed by the Bush Administration, was often accused of taking a more ideological than scientific perspective." *Nature*'s news blog (at http://blogs.nature.com), which covered the disbanding under the title "President's Bioethics Council Disbanded" (at http://blogs.nature.com) made a weaker version of the point: "Its first chair, Leon Kass, drew fire for what some regarded as ideological decisions."

104. See Guttman and Thompson 2004.

105. From the vast literature on the foundations, strengths, and limits of deliberative democracy, see, as representative of different strands that are important in deliberative democratic bioethics, the following: John Rawls' famous thought experiment on fairness, empathy, and deliberation (2001); Jürgen Habermas' process route to deliberative democracy (1996; criteria for inclusive, non-coercive, equal deliberation are listed on p. 305); Cohen (1996) on procedure and substance; Benhabib (1996) on the link between legitimacy and deliberation; Mouffe (1999) on the ineliminable role of dissent; Dryzek (2005) on deliberative democratic consensus seeking in divided societies; and Young (2000) on inclusion and substantive democracy.

106. Dodds and Thomson (2006) assess various kinds of national bioethical organizations according to their meeting the challenges of pluralist liberal democracies, and the role of a process-deliberative democracy in this. They distinguish advisory from policy-making bodies, finding the former better at deliberative depth and the latter better at making a real-world impact; they also distinguish expert from lay/public inclusive bodies or processes, with the former being better at advocating but less representative than the latter. They define three approaches, "interest group pluralism," "interest group pluralism with public consultation," (which is the kind practiced in California in the aftermath of Proposition 71); and their own position of "contested deliberation."

107. For an account of the European, and especially the British, investment in public participation in science, its elements of public trust, the "innocent" citizen, and the search for social consensus, as well as what it has to say about the state

of science–society relations today, see Irwin 2006. On how the public gets enrolled via the "deficit theory of public groups operating in a knowledge vacuum," see Irwin 2001.

108. For a discussion and an analysis of the popularity of models for the public engagement in science and technology (which the authors term PEST), and the ways in which it falls short substantively of its democratic rationale and promise, see Demeritt, Dyer, and Millington 2009. On efforts to include public voices in bioethics, see Schicktanz, Schweda, and Wynne 2011.

109. A part of the rationale for public engagement is to deal with the problem of the authority granted to scientists, especially those who make life and death decisions or who deal with technologies that have large amounts of risk attached. Sheila Jasanoff has theorized this aspect of public participation as a bundle of "technologies of humility" (Jasanoff 2003). This resonates with strands in feminist theory calling for "modest witnessing" in science—see, e.g., Haraway 1997. For a feminist liberal theoretical treatment of the undemocratic and adversarial nature of interest group representatives on ethics committees, see Young 1990. On the inclusion of groups subject to historic injustice see Ivison 2002. Ivison shows the need for representation of "the constellation of discourses and registers present in the public sphere," as well as for iterative and open-ended decision making, all of which are largely missing from the ethical deliberation around stem cell research in which I have taken part.

110. Reports were produced on synthetic biology, cross-border research, STD research in Guatemala in the 1940s, and protecting human subjects in research (Presidential Commission on Bioethics 2010, 2011a, 2011b, 2011c).

111. Blom and de Vries (2007) suggest ways to implement deliberative democratic bioethics in conditions of internal (post)colonialism, through a discussion of the impact of the Havasupai case against the University of Arizona (*Havasupai Tribe v. Arizona State University Board of Regents*, 2010); in this article the authors discuss the plea that researchers were simply doing "good science." See also de Vries et al. 2010 on the value of including surrogates in deliberative democratic bioethics in cases of dementia; and Hussein 2009 on the potential for the co-construction of democratic infrastructure and bioethics in "developing countries."

112. See Shapin 2008 and Herzig 2005.

113. See Rabinow 2007 for an anthropology of the contemporary that pries open a space for a thinking populace, a space where questions about what the good life might be in a time of biotechnology can be and must be asked as a matter of intense intellectual and moral urgency.

114. See chapter 5 below.

115. Kristin Luker captured the complex and entrenched relations between social structural position and values behind differing women's views on abortion in the U.S. "Their position on abortion," she writes, "is the 'tip of the iceberg,' a short-

hand way of supporting and proclaiming not only a complex set of values but a given set of social resources as well." (1984, chapter 8, p. 202)

CHAPTER 3

1. Thomson et al. 1998.

2. Watson and Crick 1953. This paper is archived by *Nature* with other papers from the "*annus mirabilis* for science," including Franklin and Gosling 1953, at http://www.nature.com.

3. Yamanaka and colleagues published their iPSC success in *Cell* (Takahashi et al. 2007) almost simultaneously with Thomson and colleagues' publication of their results in *Science* (Yu et al. 2007). The mouse iPSC paper came out a year previously (Takahashi and Yamanaka 2006).

4. The text of the speech given by President Bush on August 9, 2001 at Crawford, Texas is available at http://georgewbush-whitehouse.archives.gov/news/releases/2001/08/20010809-2.html.

5. The text of the remarks of President Obama prepared for the signing of the Stem Cell Executive Order and the Scientific Integrity Presidential Memorandum is available at http://www.whitehouse.gov/the_press_office/Remarks-of-the-President-As-Prepared-for-Delivery-Signing-of-Stem-Cell-Executive-Order-and-Scientific-Integrity-Presidential-Memorandum/. Excerpts from text of the Executive Order "Removing Barriers to Responsible Scientific Research Involving Human Stem Cells" are available at http://www.whitehouse.gov/the_press_office/Removing-Barriers-to-Responsible-Scientific-Research-Involving-Human-Stem-Cells/. The text of the "Memorandum for the Heads of Executive Departments and Agencies on the subject of Scientific Integrity" is available at http://www.whitehouse.gov/the_press_office/Memorandum-for-the-Heads-of-Executive-Departments-and-Agencies-3-9-09/. In this section, words in quotation marks and attributed to Obama are drawn from these texts.

6. This is what I am undertaking in this book.

7. This position has been called "threshold deontology"; see, e.g., Alexander and Moore 2008.

8. While compatibilism is generally taken to be a position the reconciles free will and determinacy (see, e.g., McKenna 2009), it is not uncommonly used to refer to compatibility between science and religion and/or ethics, especially in the U.S. Maienschein (2007) lays out the parallels between the human embryonic stem cell debate and the debate over intelligent design in the U.S.

9. Consequentialism, like deontology and compatibilism, comes in many varieties, but is generally taken to refer to ethical positions that give paramount importance to the total good in the world and the outcomes of actions, rather than to duty or principle or motive. See, e.g., Sinnott-Armstrong 2011.

10. And it is in these kinds of differences between ethical positions and ethical practice that the value of an empirical, though not solely local, approach to ethics reveals itself.

11. On the perceived strengths and weaknesses of deontological ethics, see Alexander and Moore 2008. On the perceived strengths and weaknesses of consequentialism, see Sinnott-Armstrong 2011.

12. Although the majority of Americans polled have been in favor of making abortion legal in some but not all cases since the 1970s, 2009 was the first time in over a decade since an absolute majority chose the label "pro-life" over "pro-choice" (Saad 2009). Applying the logic of "majoritarian consensus," to the question of abortion, Obama should perhaps have worked on both topics together, moving to make abortion legal only in some specific circumstances.

13. For the exact wording, visit http://stemcells.nih.gov.

14. It is not uncommon to geographically displace, or "outsource" or "off-shore," embryo politics and ethics; both Germany and Switzerland, for example, have restricted human embryonic stem cell derivation at home while allowing embryonic stem cell lines to be imported from elsewhere; and many kinds of cross-border reproductive care, from abortion seeking to third-party reproductive technologies, rely on just this kind of ethical gradient when embryo politics limit access to the procedures at home. Bush's novelty was to restrict embryo destruction in time rather than place—to before August 9, 2001—and to the private sphere.

15. The full text is available at http://georgewbush-whitehouse.archives.gov/news/releases/2007/06/20070620-6.html.

16. It should not be assumed that restrictions on procurement of material for scientific research necessarily stymie creativity and innovation if other conditions are right; it can even be its own kind of spur to innovation; see Thompson 2007a.

17. See, e.g., Stadtfeld et al. 2008.

18. See, e.g., Chen 2009.

19. See, e.g., ABC News–Washington Post Poll, July 26–30, 2001 (N = 1,352 adults nationwide; margin of error 2.5). This poll showed 63% in favor of embryonic stem cell research in July 2001, and 33% opposed, with 60% in favor of the government providing funds for the research and 36% opposed. In the days immediately following the August 9. 2001 speech, the Ipsos-Reid Poll, for example, showed high levels of interest from both parties (53% of Republicans and 52% of Democrats in the categories of "extremely" or "somewhat closely" following the debate), strong support for hESC with unused IVF embryos (71% of Republicans and 80% of Democrats), and strong support from Republicans (76%) and a majority support from Democrats (51%) for the president's compromise on federal funding (Ipsos-Reid Poll, August 10–12, 2001. N = 1,000, margin of error 3.1).

Summaries of U.S. polls on stem cell research are available at http://www.pollin-greport.com.

20. See http://www.whitehouse.gov/the_press_office/Removing-Barriers-to-Responsible-Scientific-Research-Involving-Human-Stem-Cells/ and http://www.whitehouse.gov/the_press_office/Memorandum-for-the-Heads-of-Executive *chapter 3*-Departments-and-Agencies-3-9-09/.

21. Consolidated Appropriations Act, 2009, Pub. L. 110–161, 3/11/09, known as the Dickey-Wicker Amendment.

22. See chapter 5 for a discussion of chimeras and cybrids. See chapter 8 of Thompson 2005 for a discussion of "too much hybridity."

23. For example, Obama's position accorded with the majority of the recommendations of the Center for Genetics and Society.

24. See Thompson 2007b, 2008, 2010b. See also Strathern 2011 for a discussion of my position and that of others especially in the North American context who have begun to argue for the compatibility of gift and recompense. See also Nuffield Council on Bioethics 2011, a report resulting from a public consultation carried out by a working party chaired by Dame Strathern (I provided comments), which noted the broad support for altruism, though potentially combined with certain kinds of recompense, underlying beliefs about acceptable donation of human body parts in the U.K.

25. *Sherley v. Sebelius*, 2010 U.S. Dist. (D. D. C. August 23, 2010).

26. *Sherly v. Sebelius*, 686 F. Supp. 2d 1 (D. D. C. 2009); *Sherley v. Sebelius*, 610 F. 3d 69 (D. C. Cir. 2010).

27. The entire text of Proposition 71 is available at http://www.cirm.ca.gov/sites/default/files/files/about_cirm/prop71.pdf.

28. For example, a book on the birth of Dolly the Sheep defined pluripotency as "between the extremes of totipotency and specialization" (Wilmut, Campbell, and Tudge 2000, p. 53), mostly using "pluripotent" to refer to what subsequently became known as "adult stem cells." Wilmut et al. referred to "cloning by nuclear transfer" (p. 56), and found in the successful birth of animals by somatic cell nuclear transfer the evidence that the "ICM [inner cell mass] nuclei that gave rise to them must indeed have been totipotent." (p. 114).

29. See below. Also see Dickenson 2006 on the missing women on both sides (the embryo-right-to-life side and the pro-science side) of the stem cell debate, Waldby 2008 on egg donation as gendered reproductive labor, Waldby and Cooper 2010 on egg donation for stem cell research as "regenerative labour" of the female body, and Thompson 2005 on the "biomedical mode of reproduction."

30. On the Black Panther health-care movement, see Nelson 2011.

31. Francine Coeytaux, a former Associate at the Population Council in New York, founded the Pacific Institute for Women's Health. Before the passage of

Proposition 71, she served on California's Advisory Committee on Human Cloning from 1999 to 2003. She testified for the Pro-Choice Alliance for Responsible Research at the Joint Oversight Hearing on the Implementation of Proposition 71 on March 9, 2005, before the California Senate Subcommittee on Stem Cell Research Oversight, the Senate Health Committee, and the Assembly Health Committee.

32. For the perspective drawn from her legal and policy work, see Charo 2005.

33. On the politics of voice, see chapter 1 above.

34. See, e.g., Mason, Goulden, and Frasch 2011.

35. Although I published an early version of this part of this chapter, briefly addressing the issue (Thompson 2008).

36. For example, the Ethnic Chambers of Commerce were involved, and questions of the biologizations of race and racializations of stem cell biology (see Thompson 2006), of access and affordability of treatments, and of the classism of the ICOC and its preference for elite grantees preoccupied several speakers.

37. On how religious views make their way into biomedical practice in the U.S. whether they are explicitly invoked in policy or not, see Thompson 2005 and Thompson 2007c.

38. See, e.g., Parens and Asch 2000.

39. On the ways in which race, gender, sexuality, class, disability, and immigration status have continued to be enacted through and in turn produce U.S. conceptions of reproductive privacy in the time period of this book: on surveillance, race, sexuality, and parallels and disjunctions between reproductive and sexual rights, see Eng 2010 and Miller 2000. On reproductive justice, privacy, race, citizenship, and class, see Roberts 1999, Silliman, Fried, Ross, and Gutierrez 2004, and Goodwin 2011. See also the landmark U.S. Supreme Court case *Lawrence v. Texas*, 539 U.S. 558 (2003), in which the court held that intimate consensual sexual conduct was protected under the due process clause of the Fourteenth Amendment (with parallels to the right to abortion and to the politics of race), and effectively made same-sex activity legal in every state. See also H.R. 1822: Susan B. Anthony and Frederick Douglass Prenatal Nondiscrimination Act of 2009, sponsored by Congressman Trent Franks (R-Arizona). It proposed "to prohibit discrimination against the unborn on the basis of sex or race"; see Jesudason 2011. Finally, see Stern 2005 on the history of science, sex, gender, race, and eugenics in California.

40. In the run-up to the 2012 U.S. presidential election, Republican primary candidate Newt Gingrich made clear that stem cell research had become integral to the abortion debate by pledging to ban hESC and review IVF regulation, should he be elected.

41. See Franklin and Roberts 2006.

42. For a medical perspective on the relevance of the South Korean debacle to egg donation, see Steinbrook 2006.

43. At the national level, Judy Norsigian of the Boston Women's Health Book Collective alerted women's-health advocates to the risks involved in egg donation for stem cell research (Norsigian 2005), and testified before the U.S. House of Representatives Committee on "embryonic stem cell research after Seoul."

44. The National Academies embryonic stem cell research guidelines were published earlier that year, National Research Council 2005 (amended in 2007, 2008, and 2010).

45. The full text of the letter is available at http://www.geneticsandsociety.org.

46. See, e.g., Stevens and Beeson 2006. Feminist advocates faced threats of lawsuits during the campaign for Proposition 71 when they claimed that SCNT would need thousands of women's eggs, and that it was a form of cloning.

47. On pressure to donate eggs, and on the naturalization of egg donation, see Thompson 2005 (chapter 5) and Thompson 2009.

48. See Almeling 2011 for the salience of evidencing the right emotional reasons for "donating" eggs (but not sperm).

49. See, e.g., Arditti, Duelli-Klein, and Minden 1989 (an early anthology).

50. Pressure to donate can be felt for other than financial reasons, ranging from preferred access to treatment, or the pull of family or friendship ties, to the medical and other attention given to donors, or the "foot in the door" that donation with cross-border travel can give to immigrant donors. It has been recognized, for example, that egg sharing in countries where this is offered, because it brings down the cost of, and thus access to, treatment, exerts another kind of pressure that might lead people to discount the negatives of donation. Empirical studies, however, have suggested that a viable option in a constrained field of choice better describes most patients' experiences of egg sharing than exploitation. For patients' experiences of egg sharing in the U.K., see Blyth 2004 and Haimes, Taylor, and Turkmendag 2012. See also Ethics Committee of the American Society for Reproductive Medicine 2003.

51. Ethics Committee of the American Society for Reproductive Medicine 2000.

52. Ethics Committee of the American Society for Reproductive Medicine 2007.

53. On the shift from other kinds of "reproductive work" to "regenerative labour," see Waldby and Cooper 2010. On the economic significance of adding all kinds of reproductive labor to productive labor in theories of capitalism, see chapter 8 of Thompson 2005.

54. *Eggsploitation* is the title of a 2010 documentary, directed by Jennifer Lahl, that follows the untold stories of egg donors in the fertility industry.

55. For different perspectives on circulations and alienations of different body parts in contemporary Anglophone culture, see, e.g., Waldby and Mitchell 2006; Lock

2001; Andrews and Nelkin 2001; Konrad 2005; Washington 2011; Skloot 2010. For a discussion of egg donation for research that asks how it differs from other kinds of research donation, rather than from ARTs, see Magnus and Cho 2005.

56. On why we should think about the two contexts together, see Ikemoto 2009.

57. For some of the legacies of the racialized nature of reproductive logics, see Roberts 2003.

58. On recent ways of thinking about commodification in the light of biomedical and other aspects of globalization, see Scheper-Hughes and Waquant 2002. On the notions of coercion, commercialization, and commodification in egg donation for stem cell research, see Rao 2006.

59. Thompson 2007b.

60. The Family Caregiver Alliance notes that more women than men live with disabilities (due in large part to the longer life expectancy of women), and that studies have estimated that between 59% and 79% of family and informal caregivers are women in the U.S., the minority women taking on even more of this care than white women as a result of inability to afford paid care (source: http://www.caregiver.org). The National Organization for Women reports that 90% of paid care workers are women (source: http://www.now.org). On the racialized, gendered history of coercive care work in the U.S., see Glenn 2010.

61. See Rapp 2000 on the gendered nature of these "moral pioneers." See also Taussig 2010 on the ways care work includes not just physical and emotional work but also requires cultural work, in the case of Dutch genomics, making genetic diagnoses line up with acceptable ideas of citizenship (ordinariness and tolerance in a changing Netherlands).

62. See, e.g., the work of the groups mentioned above in this chapter, namely, the Pro-Choice Alliance Against Proposition 71, the Center for Genetics and Society, Hands Off Our Ovaries, Generations Ahead, and Sins Invalid.

63. CIRM Regulations Chapter 2. Scientific and Medical Accountability Standards. Section 100110. Fairness and Diversity in Research, and CIRM Regulations Chapter 4. Intellectual Property and Revenue Sharing Requirements for For-Profit Organizations. Section 100407 Access Requirements for Products Developed by For-Profit Grantees,.

64. This format for CIRM press releases began in late 2008. The first such I received, in October 2008, recorded spending to date totals as follows: "To date, the CIRM governing board has approved 229 research and facility grants totaling more than $614 million"; this already made "CIRM the largest source of funding for human embryonic stem cell research in the world."

65. *The New Yorker* ran an article about the laboratory in September 2011, Goldberger 2011.

66. J. Craig Venter, "New Frontiers of Genomic Research," 2011 CIRM Grantee Meeting. The speech is available at http://www.cirm.ca.gov. Venter described CIRM as the "greatest scientific organization founded in our era."

67. The monthly top research picks of CIRM's president, Alan Trounson, are available at http://www.cirm.ca.gov.

CHAPTER 4

1. For coverage of the brain drain from the U.S., see Kahn 2001 for the Bush policy push factor and Watt 2006 for the EU pull factor after it increased funding. See Wolinsky 2009 for a discussion of the role of the Bush decision in a longer history of scientific brain drains in and out of the U.S. It is not only in the U.S. that "gaps" in funding for and legality of stem cell research have been conceptualized as provoking a brain drain; see Krones et al. 2008 on Germany and Dolgin 2010 on Australia.

2. Roger Pedersen was the "poster child" for the brain drain when he left UC San Francisco in 2001. The University of Cambridge lured him with more than a million pounds of government funding from Britain's Medical Research Council International Appointments Initiative. While this was no doubt a major incentive in and of itself, Pedersen was quoted as saying that he wanted to work where there was "public support," and that the separation of private and public funds, only the former of which could be used for embryonic stem cell research, had become untenable at UC San Francisco when he was told that government-funded basic lab facilities like lighting and electricity could not be used for hESC research either. (See, e.g., Highfield 2001.) The two couples who went to Singapore in 2006, Judith Swain and Edward Holmes from UCSD and Nancy Jenkins and Neal Copeland from the NIH, were part of the second rhetorical phase of the U.S. stem cell brain drain to "the East" that I discuss below. These scientists were probably pulled by the facilities and the high salaries for foreigners as much as pushed by restrictions at home, especially the scientists from California where CIRM monies were by then available; as Hamilton (2007) opined, by this point California was already benefiting from a "brain gain," having lured Shinya Yamanaka to the Gladstone Institute. Likewise, Singapore had already lured prominent U.S. life-sciences researchers, such as the cancer researcher Edwin Liu Tak-Bun in 2001, not in stem cell research and before Bush's restrictions on funding.

3. An actual loss of competitiveness and the fear of one were not the same thing. Assessments of the U.S. as leading the international competition during this period nonetheless continued to worry about the long term damage to U.S. leadership from internal policy problems; see Salter and Harvey 2008. Sipp (2009) failed to find that Asia overtook the U.S. in stem cell research competitiveness as indicated by research infrastructure and governance.

4. Some Europeans worried that the previous supposed brain drain would be reversed after Obama's lifting of Bush's federal funding restrictions; see Trager 2009. The possible threat to international competitiveness came up again with the 2011 EU ban on stem cell patents. Commentators on the hESC patent landscape differed on whether this would prompt business and/or researchers to move from Europe to Asia and the U.S. See, e.g., Plomer 2011 and Morrison 2011.

5. The full text is available at http://www.whitehouse.gov/the_press_office/Removing-Barriers-to-Responsible-Scientific-Research-Involving-Human-Stem-Cells/.

6. SB 253, sponsored by Deborah Ortiz, Democrat from Sacramento, altered the California Health and Safety Code § 123440, 24185, 12115–7, 125300–320.

7. California Governor Davis announced that "with world-class universities, top-flight researchers and a thriving biomedical industry, California is perfectly positioned to be a world leader in this area" (quoted in Martin 2002).

8. In a typical story, the magazine *New Scientist* reported that the bill's authors hoped that the bill would "encourage leading U.S. stem cell researchers to remain in or move to California" (Young 2002).

9. Johnson and Williams (2006) discuss the various state stem cell initiatives in this period in terms of regulatory arbitrage and brain drain, in a report for Congress. In the summary to their report they wrote: "federal policy has limited federal funding for research on embryonic stem cells. In response, many states are moving forward with their own initiatives to encourage or provide funding for stem cell research in order to remain competitive and prevent the relocation of scientists and biotechnology firms to other states or overseas."

10. See Brush 2005, which characterized all the U.S. state-based stem cell research funding initiatives as efforts to avoid brain drain.

11. Hwang et al. 2004, 2005.

12. Quoted in Garvey 2005.

13. In addition to the articles discussed here, see Holden 2005 and Cherwin 2007.

14. Einhorn, Veale, and Kripalani 2005.

15. For a timeline of the controversy, see *Nature*'s December 2005 editorial (available at http://www.nature.com). For more background, see Kennedy 2006 and *Science*'s "Special Online Collection: Hwang et al. Controversy" (available at http://www.sciencemag.org).

16. Somers 2006.

17. See Owen-Smith and McCormick 2006 for one of the few attempts to measure the impact of the Bush policy on the scientific competitiveness of the U.S.; they found that the number of published papers on hESC research rose significantly in the U.S., but dropped sharply as a proportion of the worldwide total between 2002 and 2004. See Korobkin 2006, for caveats in interpreting this

finding, including the fact that U.S. scientists were still receiving more support from all sources combined for hESC research than scientists anywhere else.

18. See for example, University of California 2010. Although it is difficult to read this data (international students are aggregated; permanent residents, regardless of citizenship, are counted with citizens; and Asian American ethnicity is a complex category) it is clear that California and the Asian Pacific countries had strong links in medicine and the life sciences at the graduate level throughout this period.

19. This formulation is drawn from the argument presented by a visiting lecturer to my university in 2005 who showed a power-point slide of a statue of Buddha while discussing the advantageous regulatory environment in Asia, and then a slide representing Christianity while discussing U.S. federal funding restrictions. For an attempt to put European Green arguments against hESC into the same boat as Catholic and Evangelical objections to the research as Western spiritual arguments, against an Asia not so constrained by "the God Effect," see Silver 2004; see also Tierney 2007. These kinds of accounts ignore the importance of Christianity in Asia; conflate Eastern religions; ignore the political and regulatory aspects of religion such as: efforts on the part of some East Asian stem cell ethicists to come up with Confucian ethical precepts to guide stem cell regulation (see Liu 2008), and politically enforced Confucian revivalism (see Cho 1998); equate North American and European progressive views against tampering with nature with a post-Christianity that reveres Nature as God despite the fact that most adherents of these views use non-religious idioms; fails to distinguish between abortion politics in Catholicism and in Evangelical Protestantism; ignores non-religious or differently religious reasons for supporting or opposing stem cell research; and ignores political, legal, economic, and scientific reasons for regulatory gradients for stem cell research.

20. While I learned a lot about the brain-drain imaginary from this part of my research that I could not have learned staying in the U.S., the limits of this work in ethnographic terms is self-evident: my trips were short and my Korean language skills non-existent. The global aspects of scientific knowledge and its international social organization somewhat facilitated access and comprehension, and the dominance of the English language as the lingua franca of scientific research and as a first or a second language around the world reduced linguistic barriers, as did the growing numbers of English-speaking social science scholars.

21. For Asian-nation-specific ethnographic accounts of genetic and stem cell science in which the scientific practice embeds national concerns, see Liu 2008 and Ong 2005. For an ethnography of genetics from a European country, showing that this is not an Asian (or non-Western or non-scientific) phenomenon, see Taussig 2009.

22. On Singapore's approach to hESC at the time, see Kian and Leng 2005; on its Bioethics Advisory Committee, see Ho, Capps, and Voo 2010. See Jung 2010

for an assessment of the revision after Hwang's fall of the Korean Bioethics and Biosafety Act, originally enacted in 2005.

23. See Rei and Tai 2010 on the similarities and differences among nations in governing East Asian biotechnology.

24. On the "multilateral and polycentric, though not completely multidirectional" migration of skills in "the new geopolitics of knowledge," see Meyer, Kaplan, and Charum 2001.

25. On the idea of the "developmental state" in East Asia and elsewhere, see Woo Cummings 1999. For the argument that the developmental state "incubates" conditions of scientific fraud in Northeast Asia applied to the life sciences, including the case of Hwang in Korea, see Lee and Shrank 2010. For the argument that scientific fraud and misconduct is exacerbated by conditions governing science in the U.S., on the other hand, see LaFollette 2000.

26. The United Nations Human Development Index is calculated using a combined measure of education, life expectancy, and GDP, whereas the IMF "advanced economy" designation uses the macroeconomic criteria of per capita income, export diversification, and integration into the global financial system. By 2011 (with a slightly different methodology), South Korea stood at 15, and Singapore at 26. Both fall back, especially Singapore, when inequality is factored in. Data are available at http://hdr.undp.org.

27. For a defense and exemplification of transnational comparative work, see Jasanoff 2005.

28. See Hong 2008 for an account of the Hwang scandal. For background to Korean biotechnology governance against which this took place, see Yoon, Cho, and Jung 2010.

29. See Leem and Park 2008 for the relations of Korean women to Hwang's rise and fall. For a thought experiment on how the revised Korean bioethics law and the U.S. National Academies Guidelines on Human Embryonic Stem Cell Research would have protected egg donors during the Hwang scandal, see Kim 2009.

30. For this research, given its time frame and given that this is emphatically not a book about scientists, I opted to work with conditions of anonymity, and thus do not give identifying information for those who graciously showed me around labs and talked to me.

31. As an advocate for the payment under some circumstances of women for donating eggs for research (e.g., Thompson 2007b), I was less concerned here as elsewhere by the charges that some egg donors had been paid than by the evident lack of consultation of women, possible coercion of junior colleagues, and the misrepresentation of the numbers of eggs used. A major aspect of the scientific breakthrough attributed to Hwang and his team was the ability to reduce the

numbers of eggs used to succeed in cloning. On ethical issues having to do with waste and profligacy in cloning, see pp. 245–276 of Thompson 2005.

32. On the historical and contemporary associations between given locations, particular animals, and kinds of research see Franklin 2007.

33. On scientific fraud as a spectrum continuous with normal science with its pressures, mistakes, ambiguity, and so on, and on the extent to which this means that fraud is discovered where and when it is looked for and named as such, see Martin 1992.

34. University of Pittsburgh Summary Investigative Report on Allegations of Possible Scientific Misconduct on the Part of Gerald P. Schatten, Ph.D.

35. See Ong 2005 on the nature of Singapore as a hub; and Parayil 2005 and Kao 2009 on Biopolis as an example of Singapore's new knowledge society approach to innovation in the life sciences.

36. See Smaglik 2003 on the luring of international science stars to this research real estate.

37. http://www.a-star.edu.sg/?tabid=860

38. http://www.a-star.edu.sg/AboutASTAR/Overview/tabid/140/Default.aspx

39. On the relations between space and Singapore's "new economy," see Wong and Bunnell 2006. Waldby (2009, p. 368) characterizes Biopolis' role in One North as a "garden of innovative Eden."

40. For an account of Singapore's brain-drain conditions as a combination of social authoritarianism—for example, the ban on most kinds of chewing gum in the city state—with its quest for biomedical freedom, see Arnold 2006. For an account that frames Singapore's investment in biomedicine as a project in liberal democracy, see Holden and Demeritt 2008.

41. See Ong 2007 for the idea of the pied-a-terre citizen and a global meritocracy in Asian cities; Biopolis is one of her case studies.

42. See You and Korzh 2005, written by researchers from Proteos about the establishment of zebrafish colonies in Biopolis.

43. See http://www.bioethics-singapore.org/.

44. Korea was moving in other sectors strongly in a knowledge society direction. Park, Hong, and Leydesdorff (2005) argued that South Korea had strong knowledge society credentials, but that its biomedical sector lagged behind.

45. The ministry's English-language website (http://www.mke.go.kr) says this: "In 2008, with the launch of the administration of President Lee Myung-bak, the Ministry of Knowledge Economy (MKE) was born. Formerly the Ministry of Commerce, Industry and Energy, MKE incorporates certain functions that were previously the responsibility of other Ministries (Information and Communications, Science and Technology, Finance and Economy). MKE is making painstaking

efforts to develop Korea into a knowledge-based economy, one that is driven by technological innovation." (Source: http://www.mke.go.kr)

46. For a quantitative and qualitative assessment of recent changes in Korea's demographics, government, and their relations to its innovation system, see ESCAP 2011, pp. 149–162.

47. Quoted in Faiola 2006.

48. By 2006, Singapore was ranked "the easiest place in the world to do business" by the International Finance Corporation and the World Bank, a rank that it has kept since then. The rankings are available at http://www.doingbusiness.org.

49. The United States-Singapore Free Trade Agreement of 2003 was implemented in both countries on January 1, 2004, so was in force at the time of my visit.

50. The Republic of Korea–United States Free Trade Agreement known as KORUS FTA was signed on June 30, 2007. The renegotiated FTA was signed in December 2010, and was finally ratified in late 2011 by both countries. This agreement was hotly contested in the U.S. over automobiles; during my second visit to Korea in 2008, the politics in Korea of beef imports from the U.S. sparked the Candlelight Protests against President Lee Myung-Bak and trade policies with the U.S. On its relevance to life-sciences industries, intellectual property, and patents, and on some items of potential relevance to stem cell research IP on which differences remain, see Corless, Kim, Park, and Kim 2011.

51. Overland 2007.

52. Available at http://tomorrow.sg.

53. Lee 2006.

54. Colman joined King's College in May 2008; he left in September 2009, feeling that both places deserved more of his time, and concluding that resources and working conditions were better in Singapore. Between December 2009 and November 2011, he had an adjunct position with NUI Galway, Ireland, that involved several visits there, while serving as the Executive Director of the Singapore Stem Cell Consortium. Personal communication, April 2012.

55. For the emergence of the "global biological" as an increasingly important form of global capital and a new kind of biological, of which stem cell research is a vital part, see Franklin 2005. For a EU-based perspective on the interrelations between the national and global in worldwide stem cell research and its politics, see Gottweis, Salter, and Waldby 2009.

56. Available at http://www.cirm.ca.gov.

57. CIRM Press Release, July 6, 2009, on new NIH Guidelines.

58. Diabetes is the disease from which the architect of Proposition 71, property developer Robert Klein's son suffers. The connections between diabetes, the Juvenile Diabetes Research Foundation, and human embryonic stem cell research in the U.S. began early. Harvard University's Douglas Melton, also the father to two

children suffering from juvenile diabetes, announced in early 2004 that he had derived new hESC lines with private funds provided by Harvard, the Howard Hughes Medical Institute, and the Juvenile Diabetes Research Foundation.

59. Another explicitly international body in the field is the International Stem Cell Forum, a network of 21 stem cell funders whose explicit aim is "to encourage international collaboration and funding support for stem cell research, with the overall aim of promoting global good practice and accelerating progress in this vitally important area of biomedical science" (source: http://www.stemcellforum.org). All but two of its members are government-funded national level research organizations. The two exceptions are the wealthy patient-advocate organization, the U.S. Juvenile Diabetes Research Foundation, and the California Institute for Regenerative Medicine, CIRM.

60. Among the disease advocate and government funders of the ISSCR's *Guidelines* report, EuroSTELLS and the European Commission, Sixth Framework Programme imparted the European Union's internationalizing mandate into the work. The framework's requirements for collaboration between EU countries with more and less stem cell research infrastructure in its grants drove this internationalization.

61. See Chen 2009 and Bharadwaj and Glasner 2009 for views based on empirical work in China and India (respectively) on the particular and more general challenges of stem cell tourism and the counter argument to what Chen calls the "Wild East" portrayal.

62. ISSCR, *Guidelines for the Clinical Translation of Stem Cells.*

63. The task force was co-chaired by Insoo Hyun, one of the authors of the paper on informed consent in Hwang's experiments published in the *American Journal of Bioethics* that had been retracted after the allegations about egg buying and coercion surfaced. See below.

64. See Thompson 2011 and Franklin 2011.

65. ISSCR, *Guidelines*, p. 4.

66. Federal Trade Commission "At-Home Genetic Tests: A Health Dose of Skepticism May Be the Best Prescription," at http://www.ftc.gov For the FDA regulation of DTC genetic tests, see the Genetics and Public Policy Center site, http://www.consumer.ftc.gov.

67. On the emerging health relevant DTC genetic testing landscape in 2008, see Geransar and Einsiedel 2008.

68. Crucial to the regulatory difficulties facing regenerative medicine in the future are its basis in tissue and cell transplantation, and the fact that a large proportion of transplants cross international borders. See AHCTA, 2008. "Towards a Global Standard for Donation, Collection, Testing, Processing, Storage, and Distribution of Allogeneic HSC and Related Cellular Therapies," available at http://www.ahcta.org.

69. See Deloitte Center for Health Solutions 2008 for a report on and analysis of the take off of medical tourism markets.

70. American Medical Association, 2008. "New AMA Guidelines on Medical Tourism," AMA. Available at http://www.ama-assn.org.

71. ISSCR, *Guidelines*, p. 4.

72. On the willingness of a sample of Canadian healthy adults to imagine traveling to India, China, Russia, Costa Rica, and Argentina for stem cells should the need arise, see Einsiedel and Adamson 2012.

73. In 2011 and 2012, several special issues of biomedical and social-scientific journals took on the question of medical tourism, pointing to the haves and have-nots, it inherits, creates, alters, and/or exacerbates. Of particular relevance to the argument here, see the following special issues: *Anthropology and Medicine* 18, no. 1, 2011; Body and Society 17, no. 2–3, 2011; *Reproductive Biomedicine Online* 23, no. 7, 2011; *Developing World Bioethics* 12, no. 1, 2012.

74. See, e.g., Turner 2007 and A. Wilson 2011.

75. In 2011, the two leading American plastic surgery associations, the American Society for Aesthetic Plastic Surgery, and the American Society of Plastic Surgeons, released a joint position statement on the use of stem cells in plastic surgery. Recommendations are available at http://www.surgery.org.

76. See McGee 2006.

77. On the reform of Korea's bioethics law after the scandal, see Jung 2010.

78. For the controversy regarding the editor-in-chief of the *American Journal of Bieothics*, Glenn McGee, joining CellTex before he had left the journal, for the links between Seoul based RNL Bio and CellTex, and for the perspective of Insoo Hyun, the Case Western Reserve stem cell bioethicist and joint author of the Hwang paper on informed consent that McGee pulled 6 years previously, see Cyranoski 2012a. For AJOB's editors' perspective, see Magnus 2012. For the chain of events and background on CellTex, the company from which Governor Rick Perry of Texas received controversial stem cell treatments, as a potentially "shady" stem cell company, see Shanks 2012 and Cyranoski 2012b.

CHAPTER 5

1. Source: Farlex Inc. Financial Dictionary, 2012.

2. According to Christopher Scott (2003), "discussions of these statements typically include words such as estimate, anticipate, project, and believe."

3. The full text of Proposition 71 is available at http://www.cirm.ca.gov/sites/default/files/files/about_cirm/prop71.pdf.

4. For the concept of "pro-curial" biomedical research, see chapter 3. Its characteristics are an ideological emphasis on being *pro-cures*, an intense concern with

acceptable *procurement* of human tissue needed for the research, and an ongoing need for the *curation* of bio-information and biological material.

5. Michael Fortun and Skúli Sigurdsson have portrayed risk lists in SEC filings as the flip-side of the over-promising on the value of investment in new genetic technologies, in their work on the Icelandic government's collaboration (through a highly controversial presumed consent policy on use of the Health Sector Database and a $200 million loan guarantee) with Hoffmann-La Roche-backed U.S. venture capital company deCODE Genetics Inc., and its population genetics-based drug discovery program. While the transnational and Icelandic setting of the deCODE debacle, and Proposition 71 were very different—especially the tax-payer/voter-approved-and-funded aspect of Proposition 71, the California location, and the bioethical instruments in play—many of their arguments about biomedical promising and public-private collaborations on private wealth creation in the biotech sector resonate. See, e.g., Sigurdsson 2002 and Fortun 2008.

6. A Senior Research Scientist was a CIRM Postdoctoral Fellow before joining the company in October 2011.

7. See Galef 2011.

8. On the reasons for this, see David Jensen's California Stem Cell Report entry for November 7, 2011, available at http://californiastemcellreport.blogspot.

9. On the blurred regulatory and practical lines between eggs donated for commercial fertility treatments and egg sharing from clinics for stem cell research in California, see Reynolds 2010 and Braun and Schultz 2012.

10. The company's SEC filings are available at http://www.internationalstemcell .com.

11. The cosmetic sideline can be perused at https://www.lifelineskincare.com.

12. This company was not unique in the biotechnology sector in California in generating revenue with a sideline in cosmetics; for example, the synthetic biology company Amyris had a foray into cosmetics.

13. For cosmetic, commercial uses of pluripotent stem cells, see Harmon 2009, which assesses the ease with which fat stem cells can be reprogrammed to iPS cells, and used in medical fields including cosmetic surgery. On Cytori Therapeutics' Restore-2 breast reconstruction cell therapy trial results (likely to find cosmetic surgery markets in the U.S. if approved for clinical use there), see Pérez-Cano et al. 2012. Cosmetic surgeries are not only aesthetic. Pluristem Technologies, financed by the Israeli and other governments, is exploring the use of placental stem cells for treatment after radiation exposure.

14. This underlines the raced, gendered, and classed implications of the public support for research, as taxpayers, as providers of body parts and services, and as consumers, especially when the promise for which the research is sold to the public turns out to be very different from the eventual products marketed.

15. See "International Stem Cell Corporation Progresses towards Establishment of the Industry's First Universal Stem Cell Bank, UniStemCell" at http://www.b2i .usofiles.

16. For an analysis of egg sharing in the U.K., see Haimes, Taylor, and Turkmendag 2012.

17. I retain the language of "donation" and (occasionally) "gift" in this chapter because both words were used by people involved in the procedures. In addition to ethnographic accuracy, there are two benefits for my analysis in retaining this language. One is that the words "donation" and "gift" have powerful affective connotations that are part of the ethical choreography of biomedical donation. (See especially the section on egg donation below.) They tell us how we are supposed to act and feel, as well as naming a role and a process, in ways without which the procedures would not work smoothly or be deemed ethically acceptable. Second, the word "gift" inherits a tradition of scholarship in anthropology and social theory that provides depth to another analytic that was ethnographically significant, namely the question of reciprocity.

18. "Fiscal Impact from the Legislative Analyst," available at http://www.smart voter.org. Offsetting the cost, Alberro (2011) assesses the likely economic impact of research funded by CIRM.

19. In the U.S. in 2010, for example, an estimated 61.6% of R&D was financed by industry, with 31.3% being financed by government; 70.3% of R&D was performed by industry, with 13.5% being performed in the universities, and 11.7% being performed by other government entities, including the military (source: OECD 2012, p. 18). On the history and present of the public and private funding of science in the U.S., from different methodological and disciplinary perspectives, see Berman 2011, Mirowski 2011, and Stephan 2012. On the tortured current relationship between the academy more generally and commercialization, see Bok 2003 and Slaughter and Rhoades 2009.

20. On university-industry public funding and technology transfer to the private sector before and after the Bayh-Dole Act of 1980 (and the argument that the act was not as much of a watershed as is often suggested, and that the scale of U.S. tertiary education makes comparisons with other countries difficult), see Mowery and Sampat 2004.

21. For a more positive assessment of California's initiative process, see Baldassare and Katz 2007. For a negative view of ballot initiatives, see Broder 2001.

22. The National Science Foundation was founded in 1950 to support basic science and engineering. The National Institutes of Health trace their history much further back, but did not start investing significant sums of money, through competitive, peer-reviewed grant applications, until the early 1960s.

23. White House 2012, p. 2.

24. On the downsides of the newly intensified pressure to commercialize, see Caulfield 2012.

25. On defining and measuring the "therapeutic misconception," see Henderson et al. 2007.

26. In Cressey 2012, James Thomson is quoted as saying of the pluripotent stem cell field he began, in reference to its emerging importance in drug discovery at his company Cellular Dynamics International, the following: "'I think there are tremendous parallels to the early days of recombinant DNA in this field. . . . I don't think people appreciated what a broad-ranging tool recombinant DNA was in the middle '70s." At the same time, he says, they underestimated the difficulty of using it in treatments."

27. See CIRM 2012, p. 9–11 for CIRM's ten-year and one-year goals.

28. See chapter 2 on acceptable derivation.

29. For an account of how the attention to procurement overrode informed consent around embryo-derived stem cell lines fundable by the federal government under President George W. Bush's 2001 stem cell policy, see Streiffer 2008.

30. See chapter 3.

31. The landmark case, *Moore v. Regents of the University of California* (51 Cal. 3d 120; 271 Cal. Rptr. 146; 793 P. 2d 479), was decided by the Supreme Court of California in 1990. See also Boyle 1996 and Jasanoff 2002 for an analysis of the vital role of U.S. courts in policy making in the life sciences, and "the tendency [of the U.S. courts] to favor economic agents over those presenting merely moral claims. In Moore, for example, it is the university and its researchers who are granted property rights in the patient's excised cells, not the patient himself." (ibid., p. 897). *Moore v. Regents* has been relied upon in dismissing all or part of related cases since then; for example, in *Greenberg v. Miami Children's Hospital* (264 F. Supp. 2d 1064 (S.D. Fla. 2003)), Canavan Disease families and their disease advocacy organizations protested the commercialization of the research enabled by their tissue and monetary gifts. Charo (2006) shows that to answer the "deceptively simple" question "Do we own our bodies?" in the present-day U.S. requires asking "about our bodies and" their relationship "to our excised body tissue" and "those relationships both before and after we die."

32. On private companies' and universities' property interests in genetic samples and information, and the lack of property interests held by research donors themselves, as codified in informed-consent forms for biomedical research, see Merz, Magnus, Cho, and Kaplan 2002.

33. See e.g., Landecker 2007 and Skloot 2010. On placing several recent incidents in the context of the legal absence of property rights in our body parts, see Williams 2012.

34. For news of the agreement, see Fletcher 2010. For sample language to use in consent forms when working with Native communities, see Burhansstipanov, Bernis, Kaur, and Bernis 2005. For various approaches to informed consent, including "tiered consent," that would have dealt better with the issues at stake in the *Havasupai v. Arizona State* case, see Mello and Wolf 2010. Reardon and TallBear 2012, who place the case in a history of "whiteness as property" (Harris 1993), document Indigenous biobanking initiatives based on group consent, community property, and "DNA on loan," and suggest strategies for reframing the ethics of bio-banking so as not to replicate bio-colonialisms of the past.

35. *Beleno et al. v. Texas Department of State Health Services et al.*, Texas Western District Court, 2009; *Bearder v. State of Minnesota*, Supreme Court of Minnesota, 2011.

36. On the consequences of Proposition 69, the ballot initiative that passed at the same time as Proposition 71 in California, that authorized sweeping collection and storage of DNA of felons, and expanded DNA collection in 2009 to adults arrested for a felony regardless of subsequent conviction, see Simoncelli and Steinhardt 2006. On the links between forensic databases and the troubled history of which the new intertwinings of race and genetics are a part, see Duster 2006 and Ossorio and Duster 2005. For the implications, including the racialization, of familial DNA testing through forensic DNA databases, see Grimm 2007 and Greely, Riordan, Garrison, and Mountain 2006.

37. On the Berkeley experiment, see Thompson 2012. On the Stanford experiment, see Salari, Pizzo, and Prober 2011.

38. The Myriad case (*Association for Molecular Pathology v. Myriad Genetics*, No. 11–725 Southern District of New York, 2010) first ran counter to the famous *Diamond v. Chakrabarty*, 447 U.S. 303 (1980); in the latter the U.S. Supreme Court ruled that the question of whether an oil-consuming human-made bacterium was living or not was irrelevant to its patentability. Appeals followed the initial Myriad ruling. See also the Supreme Court 2012 decision in *Mayo Collaborative Services v. Prometheus Laboratories, Inc.*, which suggested a growing inclination to restrict the patentability of human biological systems. On the stakes in the *Myriad* case in the eyes of the biotech industry, see Wadman 2010.

39. See, e.g., Talan 2010.

40. On "named group" privacy and other concerns, see the National Institutes of Health Bioethics Resources on the Web, "Points to Consider When Planning a Genetic Study that Involves Members of Named Populations." Available at http://bioethics.od.nih.gov/named_populations.html. On individual risk, see Benitez and Malin 2010.

41. For an account and assessment of the community engagement efforts of 6 large U.S. biobanks (Group Health Cooperative, University of Washington; Kaiser Permanante, California; Marshfield Clinic, Minnesota; Northwestern University

Medical Center, Illinois; and Vanderbilt University Medical Center, Tennessee), see Lemke et al., 2010/11.

42. The journal *The Public Understanding of Science*, despite the single deficit (insufficient understanding on the part of the public) addressed in its title, claims to cover "all aspects of the inter-relationships between science (including technology and medicine) and the public" (http://pus.sagepub.com/).

43. The text of the European Union Tissue Directive is available at http://eur-lex. europa.eu.

44. These tissue types nowhere near exhaust the different frames of donation for contemporary human body parts. For example, in the U.S. breast milk is considered "food," while organs and blood and hair have their own deep histories and networks.

45. See the discussion of egg donors for stem cell research in chapter 3, and the discussion below about the affective "gift" script. See also Carson et al. 2010 for the CIRM guidelines for oocyte donation for research purposes developed with the assistance of the Institutes of Medicine and U.S. clinical experts outside California, to mitigate potential harm from ovarian hyperstimulation syndrome by excluding at risk donors, developing individualized protocols, and closely monitoring the research donor during the whole donation process.

46. On ethical issues in somatic tissue donation for stem cell research, see Hyun 2008; for the potential risks and benefits to donors and the advantages for research of re-contact, see the protocols of the major California biomedical research universities for stem cell research conducted with CIRM funding, such as UCLA's very clear checklist, available at http://www.stemcell.ucla.edu. Also see the discussion below.

47. CIRM's intellectual-property FAQs (at http://www.cirm.ca.gov) include the question "Do CIRM's regulations seek to ensure Californians pay a fair price for drugs they helped to create?" On justice, intellectual property and access, and affordability, see Bok, Schill, and Faden 2004; Winickoff 2006; and Goozner 2006. See also Tayag 2006 a and b, prepared for our conference, Toward Fair Cures, described in chapter 3. For the subsequent development of a "loophole" at CIRM, see David Jensen's *California Stem Cell Report* entry for August 18, 2009, "Consumer-Watchdog: Affordable Access to CIRM-Financed Therapies Threatened," about John Simpson of Consumer Watchdog's comments on changes to CIRM's IP policies, available at http://californiastemcellreport.blogspot.

48. Some of the more "ontological" scholars in the field of Science and Technology Studies (Thompson 2005, chapter 1) might say that there is a kind of reciprocal agency here between biological material and recipients (see Latour 1988 on microbes and Pasteur); other ontological STSers might argue that reciprocity makes no sense from the point of view of biomaterials whose biohistories and geohistories are of an altogether different scale (e.g., Myra Hird writes of a "non-

modern microontology," that the "non-human (whether biotic or not) has agency regardless of what I think," "As such, I try to move toward a sort of nonhumanism," Hird 2009, p. 20). Both of these points—biomaterials as having agency, and the fact that that agency goes much beyond the co-productive human-cellular biomaterials assemblages in stem cell research (even as being human is being redefined in "post-human" directions by regenerative medicine) are both relevant to my argument here.

49. For U.K. and other European perspectives during the period of this book on tissue banking for research, including informed consent and biobanking, and the moral frameworks of respect, reciprocity and trust, participation and solidarity, see Tutton and Corrigan 2004 and Lenk, Hoppe, Beier, and Wiesemann 2011.

50. See Rao 2005, Thompson 2007b, and Thompson 2010b for examples of arguments made against defining this relationship by donor altruism in this period. For European perspectives on the inappropriate use (because tissue for research is not a gift, does not directly help others, can be commercialized by researchers, and cannot be guided by or made confidential for the donor) and of the concept of altruism in the donation of human tissue and materials to biomedical research, and for alternatives using participation and property, see Steinmann, Sykora, and Wiesing 2009.

51. For an analysis of the notion of "undue inducement" in the Common Rule for biomedical research subjects, see Grady 2001 and Emanuel 2005; for its debate in Assisted Reproductive Technologies in the U.S., including its relevance to stem cell research, see Ethics Committee of the American Society for Reproductive Medicine 2007.

52. See http://www.bhed.com/.

53. See Blyth and Frith 2009 for jurisdictions allowing donor-conceived people's access to genetic and biographical history. British Columbia, Canada joined several European countries (United Kingdom, Norway, Netherlands, Sweden, Finland, Austria, Switzerland) and New Zealand and parts of Australia in May 2011 in banning gamete-donor anonymity, though the state chose to appeal the decision rather than amending the relevant adoption law; see Motluk 2011. This decision, part of a widespread trend toward formulating and protecting donor-conceived children's medical and identity rights to know their genetic heritage, has fueled the debate (inconclusive to date) about whether or not ending anonymity decreases the number of egg and sperm donors.

54. See Almeling 2011 on the gendered ordering of how to behave in the context of being paid for egg and sperm donation in the U.S.

55. The URL of the website through which this advice is solicited and rendered is http://askaneggdonor.com/.

56. On how these issues interrelate, see Cussins 2012.

57. For a similar "cauterizing" (his word) effect of kin relations, in a very different idiom, see Simpson 2013. Simpson describes the way that the doctor was trusted by IVF patients to choose ova donors in Sri Lankan egg donation during his fieldwork there.

58. See Weindling 2001, Rose 2007–2008, and Hoeyer 2009 on the immediate and subsequent post-World War II history and rationale of the various parts of informed consent.

59. See the full text at http://www.cirm.ca.gov. I submitted comments, just as I had done for the Nuffield Council on Bioethics report on donation for medicine and research (Nuffield Council on Bioethics 2011). My comments are available at http://www.nuffieldbioethics.org.

60. The full text of California's Research Subject's Bill of Rights is available at http://ag.ca.gov.

61. As can be seen from the fates of "abandoned" embryos cryo-preserved around the world, re-contact while protecting privacy can be fiendishly difficult. For a discussion of this in the U.S. context, in which the politics of abortion looms large, see American Society for Reproductive Medicine 2004.

62. To my knowledge, it has yet to be specified under what conditions, if any, a state-financed stem cell bio-bank might be asked to release identifying specimens and/or genetic information for forensic purposes.

63. "White Paper" on Informed Consent Protocols, cached at http://arep.med.harvard.edu.

64. See Lunshof, Chadwick, Vorhaus, and Church 2008.

65. The PGP's homepage, with links to its documentary history regarding informed consent, can be found at http://www.personalgenomes.org/.

66. Anne Wojcicki—a co-founder of 23andMe, and then the wife of Google co-founder Sergey Brin—was pregnant with their first child at the time; she referred in the SciFoo session at which she spoke to the Parkinson's in Brin's family and its relation to her unborn child's risk for the disease as a motivation for pursuing genomic research.

67. Reardon (2011, p. 100) discusses the celebrity spit parties held by companies like 23andMe as re-launching direct-to-consumer (DTC) genomics as "a thing of value to a person of taste."

68. Many who had paid several hundred dollars to have their DNA analyzed, and who had since spent hours contributing unrecompensed crowd-sourced pheno-typic, lifestyle, and medical history data to the company through its endless surveys, were shocked when 23andMe filed for a patent. Several protested on the company's website and elsewhere (examples: "23andMe has my DNA; having known that the company was seeking patents, I had been hesitant to complete their surveys, but occasionally found myself clicking through and answering, only to be

reminded of my scruples when I got a message that the survey was, e.g., '4% complete' . . . "; "I am (or my DNA and medical records is) also part of the UCSF-Kaiser genome initiative.")

69. 23andMe filed for its first patent, "Polymorphisms Related to Parkinson's Disease," in November 2010; it was published in June 2011, and U.S. Patent Number 8,187,811 issued in May 2012. See Baker 2012, which quotes co-founder Anne Wojicicki as saying that patents should be sought to "move from the realm of academic publishing to the world of impacting lives by preventing, treating or curing disease," and as believing that "patents should not be used to obstruct research or prevent people from accessing their own data." On the significance of this patent to informed consent as "a broken contract," see Hayden 2012.

70. National Research Council 2011.

71. For details on Consent to Research and the Portable Legal Consent, see its home page at http://weconsent.us/. See also "Consent 2.0," *The Economist*, April 28, 2012, and Kotz 2012.

72. On the two-year NIH study on the responsibilities of biobanks to return incidental findings, see Wolf et al. 2012.

73. For this model, a discussion of Moore, Greenberg, and the PXE International patent cases, and a comprehensive consideration of arguments for and against granting "participants' property rights in their biological material," see Gitter 2004; on the PXE case, see Smaglik 2000.

74. See, e.g., Harrison 2002 and Thompson 2007b. For the argument that these kinds of agreements would still lead to global, stratified commodification of human tissue, see Dickenson 2005.

75. The CAL-DNA Data Bank in Richmond is a laboratory that works to process and store the forensic DNA samples collected as a result of California Proposition 69, "The DNA Fingerprint, Unsolved Crime and Innocence Protection Act," that was passed by the voters at the same time as Proposition 71.

76. See, e.g., California Institute for Regenerative Medicine 2010.

77. See Chander and Sunder 2004 on "the romance of the public domain" and its challenges and possibilities regarding genomic information (which is not subject to overuse in a "tragedy of the commons" way, so does not benefit from the same rationale for propertization).

78. On notions of community, property, and benefit sharing applied to human biological material in different national and institutional public-private settings, see Hayden 2007, Parry 2005, and Høyer 2004. On group consent and questions of to whom samples belong at the time of donation and over time, see Callaway 2011 and Schrag 2006.

79. See Bellivier and Noiville 2009.

80. On CIRM's process and policies since its inception geared to "Ensuring Innovation and Fair Return," see Roth 2012. On industry distrust of CIRM's fair

return policies, see Langston 2011. On how educational and research use exemptions to certain aspects of patent restrictions or thickets create a kind of commons, see Lee 2008.

81. See, e.g., Thompson 2010a.

82. On arguments for creating a public stem cell bank, see Winickoff 2006. On a limited collective commons for open stem cell research in California, see Winickoff, Saha, and Graaf 2009. On treating stem cell research donors as partners in or allowing them to control directing sample use rather than increasing donor restrictions in the name of protection, see Saha and Hurlbut 2011 and Holm 2006.

83. My work has been described as part of a move in euro-American biomedical kinship scholarship to "re-attach reciprocity to the gift" (Strathern 2011). This is an accurate characterization in the sense that the detachment of donor altruism from any kind of benefit at the exact time (and for the same reasons) that researchers were under all kinds of pressure to benefit from that tissue—the post-Bayh-Dole, post-*Moore v. Regents* California environment that pledged public monies to cover the market failures of the bioeconomy—had increasingly come under fire, and needed addressing. Proposition 71 imagined downstream value creation and returns to the state, leaving the details around the gifts of funding and human tissue in terms of feasible reciprocity to be worked out.

84. On the notion of "digital curating" of scientific data in the context of the openness of science, see Royal Society 2012, and on curating biomaterials, see chapter 2, where I lay out the three aspects of a "pro-curial" bioeconomy.

85. I have begun working on what a portable architecture for such an initiative might look like.

CHAPTER 6

1. Loew and Cohen 2002.

2. Quoted on p. 219 of Park 2006.

3. For a summary and discussion of the laws, the literature, and opinions on markets in cadaver donation in the U.S., see Anteby 2010.

4. See Wald 2008 on biosafety, Washington 2008 on medical apartheid, and Williams 2012 on market issues around the circulation of cadavers and body parts. For a review of the qualitative literature on willingness or unwillingness to be an organ donor, and eight factors that influence that (relational ties, religious beliefs, cultural influences, family influences, body integrity, distrust of medical definition of brain death or retrieval process, knowledge about donation, and other major reservations, despite knowledge), see Irving et al. 2011.

5. The portion of research on animals that is directly related to improvements in veterinary medicine is often said to be 5 percent. See, e.g., http://www.aboutanimaltesting.co.uk.

6. Darwin was against animal experimentation for mere curiosity, as opposed to physiological investigation. See, e.g., Feller 2009 and Browne 2003.

7. For a clear exposition of the major rationales for different model organisms, see Twyman 2002.

8. On the history through the twentieth century of various model organism systems, see, e.g., Rader 2004 (mouse), Kohler 1994 (drosophila), and Creager 2001 (tobacco mosaic virus).

9. Women activists were especially important to anti-vivisectionism in the U.K. and the U.S. in the second half of the nineteenth century. On the relationship between Charles Darwin and Frances Power Cobbe, see Richards 1997. On women and anti-vivisectionism in the late-nineteenth-century U.S., see Buettinger 1997.

10. On the historical place of "the bioethical enterprise," see Rosenberg 1999.

11. On the historical origins of the Nuremberg Code, see Grodin 1992.

12. Nuremberg Code: Directives for Human Experimentation is available at http://ori.dhhs.gov.

13. On the challenge to human exceptionalism, and the interdependent subjectivities of "transspecies engagement" in "multispecies ethnography," see Kirksey and Helmreich 2010 and Kohn 2007.

14. As I, and many others, have explored, efforts to differentiate human and animal mind is of less help than might be thought in making a principled distinction between animal and human testing. See, e.g., Thompson Cussins 1999.

15. On the changing stakes over the course of the twentieth century of studying (with) animals in science, from primates to oncomouse, and of the need for companion-speciesship, see the extraordinary series of books by Donna Haraway (1989, 1997, and 2007). She "stays with the trouble" of debates such as animal testing, and does not necessarily—as I do here—advocate an end to most animal testing and use in science; nonetheless, her post-ELSI (ethical, legal, and social implications) ethics of companion species is essential to de-centering the human exceptionalism of the substitutional research subject.

16. See Till and McCulloch 1961 and Becker, McCulloch, and Till 1963. See also CIRM's blog entry on James Till's career and honors, and his thoughts on regenerative medicine today, at http://cirmresearch.blogspot.

17. Evans and Kaufman 1981.

18. Martin 1981.

19. Evans' work has received the highest recognition. The top U.S. biological sciences prize, the Lasker Award, was awarded to Evans, Mario Capecchi, and Oliver Smithies in 2001. Evans was knighted in 2004 for his services to medical science, and he received the 2007 Nobel Prize in Physiology or Medicine, also with Capecchi and Smithies, for their mouse work. The Nobel Prize was awarded for

"discoveries of principles for introducing specific gene modifications in mice by the use of embryonic stem cells" (source: http://www.nobelprize.org).

20. For an account of Evans' mouse work and his thinking, see Evans 2001.

21. This paragraph and the two quotations in the paragraph above can be found at http://www.nih.gov.

22. A registry of NIH approved embryonic stem cell lines can be found at http://grants.nih.gov.

23. *Criteria for Determining Pluripotency in Human Cells*, NIH Human Pluripotent Stem Cell Registry.

24. The website of the National Institutes of Health (http://stemcells.nih.gov) archives current and former policy documents, including the NIH Human Embryonic Stem Cell Registry, subsequently renamed the NIH Human Pluripotent Stem Cell Registry.

25. See, e.g., Xu et al. 2001.

26. Geron Corporation 2009b.

27. Akst 2009. The paper about which Zeng was talking was Zhao et al. 2009. In a further co-implication of life forms, viral genome integration from the viruses used to introduce the Yamanaka factors was a problem with these early papers, lowering successful live mouse birth rates.

28. On the limits of the human at the frontiers of biomedicine, see Squier 2005.

29. For a highly original account of the scientific and cultural contexts and significance of Dolly and of sheep as livestock more generally, see Franklin 2007.

30. Hwang et al. 2004, 2005.

31. Wilmut, Campbell, and Tudge 2000. Also see Franklin 2007.

32. Wilmut et al. 2000, p. 282.

33. Ibid., p. 287.

34. Reeve's testimony can be found at http://www.chrisreevehomepage.com.

35. The Human Fertilisation and Embryology Act of 2008 added Section 4A, "Prohibitions in connection with genetic material not of human origin," to the 1990 Act. Among other things, it specified that "No person shall . . . bring about the creation of a human admixed embryo," "admixed embryo" being the expression used for cybrids in the legislation, "except in pursuance of a license."

36. On British public opinion on cybrids, see Jones 2009. In my fieldwork, I encountered a transatlantic (and beyond) debate going on into 2009 at least, about whether cybrid embryos could even develop sufficiently to allow for the derivation of embryonic stem cell lines. In the U.S., Robert Lanza of Advanced Cell Technology argued that transcription factors Nanog, Oct4, and Sox2 were not switched on, that Chinese scientists were said to have been successful and to be

moving ahead with the research, and that some thought cybrid research to have been an embarrassing interlude best quickly forgotten. See, e.g., Ledford 2009.

37. Cressey 2009.

38. For a history of ethical concerns about chimeras and an argument rejecting these restrictions in embryonic stem cell research, see Robert 2006.

39. The bill was introduced in Congress on July 9, 2009, but was not enacted.

40. The comments are available at http://senatorsambrownback/blogspot.com.

41. For the logic and a summary of the advantages and drawbacks of different murine humanization protocols including ones using stem cells, for different aspects of HIV research, see Denton and Garcia 2009.

42. For the "three Rs," see Russell and Burch 1959.

43. See, e.g., Ito et al. 2012.

44. See, e.g., Everts 2007.

45. On the development of the argument that animal testing is bad science and the product of corporate interests rather than a necessary evil for the advancement of human medicine, see Greek and Greek 2003.

46. CIRM Initial Statement of Reasons for the Proposed Amendments of Sections 100070 and 100090, available at http://www.cirm.ca.gov.

47. CIRM Initial Statement of Reasons for the Proposed Amendments of Sections 100070 and 100090.

48. These committees review research carried out under university auspices, usually for the purposes of qualifying for federal funding; review is operationalized as compliance with relevant national and local regulation.

49. On the care and the use of animals, see Office of Laboratory Animal Welfare 2002 and CIOMS 1985. On conceptual issues having to do with human subjects and participants in biomedical research in the U.S., see Epstein 2007 and Reverby 2000.

50. Bolton 2009.

51. Sunstein 2004.

52. On standing for non-human living things, see the classic *Should Trees Have Standing?* (Stone 1974).

53. More information can be found at http://www.laanimalservices.com.

54. Milde and Yates 2009.

55. See Thompson 2002a, where I explored the relations between differing views of elephant conservation and philosophies of nature.

56. On using animal models in biodefense work, see Swearengen 2005.

57. On large-scale swine facilities and the mutual construction of nature and society, see Coppin 2002.

58. For examples of my previous work on animal geopolitics, on *in situ* and *ex situ* biodiversity conservation, and on trade in endangered species, see Thompson Cussins 1999; Thompson 2002b; Thompson 2004.

59. See, e.g., Terry 2000 and Bagemihl 2000.

60. See Jackson 2011.

61. Singer 2001.

62. See, e.g., Not Dead Yet 1999.

63. See, e.g., Hammonds and Herzig 2008, Washington 2008, Roberts 2011, and Wailoo, Nelson, and Lee 2012.

64. See, e.g., Epstein 2007 and Nelson 2011.

65. For example, the website blackvegetarians.org cites environmental, ethical (animal rights), and health benefits.

66. E.g., LaDuke 1999.

67. Mbembe 2001, pp. 1–2.

68. Biehl (2005, p. 39) shows how the poorest of the poor are (treated like) animals, and those who treat them that way—the wealthy of the world—are (acting like) animals in their medical abandonment of the poor.

69. Yudof 2008.

70. Statement of UC Chancellors on Animal Research, June 6, 2007, available at www.universityofcalifornia.edu.

71. Birgeneau 2008.

72. Jasbir Puar argues in *Terrorist Assemblages* (2007) that it is imperative to subject "the 'with us or against us' rhetoric of the war on terror" to "upheaval." 73. Nickerson 2009.

74. "GE Healthcare and Geron Announce Exclusive Global Agreement to Commercialize Stem Cell Drug Discovery Technologies," posted on the Geron website on June 30, 2009.

75. See, e.g., Gabriel 2009 and Rockstanding 2009, respectively.

76. Geron Corporation 2009b.

77. Visvanathan 1997, p. 25.

78. Ibid., p. 32.

REFERENCES

Agamben, Giorgio. 1998. *Homo Sacer*. Stanford University Press.

Ahmed, Sara. 2010. *The Promise of Happiness*. Duke University Press.

Åhrlund-Richter, Lars, Michele De Luca, Daniel Marshak, Megan Munsie, Anna Veiga, and Mahendra Rao. 2009. Isolation and production of cells suitable for human therapy: Challenges ahead. *Cell Stem Cell* 4: 20–26.

Akst, Jef. 2009. iPS cells yield live mice. *Scientist*, July 23. http://classic.the-scientist .com/blog/display/55835/.

Alarcon, Norma, Caren Kaplan, and Minoo Moallem. 1999. Introduction: *Between Woman and Nation*. In *Between Woman and Nation*, ed. C. Kaplan, N. Alarcon, and M. Moallem. Duke University Press.

Alberro, José. 2011. Economic impact of research funded by the California Institute for Regenerative Medicine. California Institute for Regenerative Medicine. http:// www.cirm.ca.gov/about-cirm/cirm-publications.

Alexander, Larry, and Michael Moore. 2008. Deontological ethics. In *The Stanford Encyclopedia of Philosophy* (fall 2008 edition), ed. Edward Zalta. http://plato.stanford. edu/archives/fall2008/entries/ethics-deontological/.

Allen, Arthur, 1997. Policing the gene machine: Can anyone control the Human Genome Project? *Lingua Franca* 7 (3): 28–37.

Almeling, Rene. 2011. *Sex Cells: The Medical Market for Eggs and Sperm*. University of California Press.

American Medical Association. 2008. *New AMA Guidelines on Medical Tourism*. http://www.ama-assn.org/ama1/pub/upload/mm/31/medicaltourism.pdf.

Andrews, Lori, and Dorothy Nelkin. 2001. *Body Bazaar: The Market for Human Tissue in the Biotechnology Age*. Crown.

Anteby, Michel, 2010. A market for human cadavers in all but name? *Economic Sociology—The European Electronic Newsletter* 11 (2): 3–7.

Arditti, Rita, Renate Duelli-Klein, and Shelley Minden. 1989. *Test Tube Women: What Future for Motherhood?* Pandora.

Arnold, Wayne. 2006. Singapore acts as haven for stem cell research. *New York Times*, August 17.

Arondekar, Anjali. 2009. *For the Record: On Sexuality and the Colonghial Archive in India*. Duke University Press.

Asch, Adrienne. 2000. Why I haven't changed my mind about prenatal diagnosis: Reflections and refinements. In *Prenatal Testing and Disability Rights*, ed. Erik Parens and Adrienne Asch. Georgetown University Press.

Bagemihl, Bruce. 2000. *Biological Exuberance: Animal Homosexuality and Natural Diversity*. Stonewall Inn Editions.

Bahadur, G., M. Morrison, and L. Machin. 2010. Beyond the "embryo question": Human embryonic stem cell ethics in the context of biomaterial donation in the UK. *Reproductive Biomedicine Online* 21 (7): 868–874.

Baillie, Harold W., and Timothy K. Casey, eds. 2004. *Is Human Nature Obsolete? Genetics, Bioengineering, and the Future of the Human Condition*. MIT Press.

Baker, Monya. 2012. Personal-genetics company patent raises hackles. Nature Newsblog, May 31. http://blogs.nature.com/news/2012/05/personal-genetics-company-patent-raises-hackles.html.

Baldassare, Mark, and Cheryl Katz. 2007. *The Coming of Age of Direct Democracy: California's Recall and Beyond*. Rowman & Littlefield.

Barad, Karen. 2007. *Meeting the Universe Halfway: Quantum Physics and the Entanglement of Matter and Meaning*. Duke University Press.

Bayliss, Françoise. 2011. A biotechnology company's decision to end an experimental stem-cell study for business considerations raises important ethical questions. *The Mark*, November 25. http://www.themarknews.com/articles/7562-where -research-ethics-meet-profit-margins.

Becker, A., E. McCulloch, and J. Till. 1963. Cytological demonstration of the clonal nature of spleen colonies derived from transplanted mouse marrow cells. *Nature* 197 (4866): 452–454.

Bellivier, Florence, and Christine Noiville. 2009. *La Bioéquité: Batailles Autour du Partage du Vivant*. Autrement.

Benhabib, Seyla. 1996. Toward a deliberative model of democratic legitimacy. In *Democracy and Difference: Contesting Boundaries of the Political*, ed. Seyla Benhabib. Princeton University Press.

Benitez, Kathleen, and Bradley Malin. 2010. Evaluating re-identification risks with respect to the HIPAA privacy rule. *Journal of the American Medical Informatics Association* 17: 169–177.

Benjamin, Ruha. 2013. *People's Science: Bodies and Rights on the Stem Cell Frontier*. Stanford University Press.

Berman, Elizabeth Popp. 2011. *Creating the Market University: How Academic Science Became an Economic Engine*. Princeton University Press.

Berne, Patty. 2008. Sins invalid: Disability, dancing, and claiming beauty. In *Telling Stories to Change the World: Global Voices on the Power of Narrative to Build Community and Make Social Justice Claims*, ed. Ricky Sollinger, Madeline Fox, and Kayhan Irani. Routledge.

Bharadwaj, Aditya, and Peter Glasner. 2009. *Local Cells, Global Science: The Proliferation of Stem Cell Technologies in India*. Routledge.

Biehl, Joao. 2005. *Vita: Life in a Zone of Social Abandonment*. University of California Press.

Birgeneau, Robert. 2008. Message from Chancellor Birgeneau: Researchers must be free from threats and violence by animal rights terrorists. http://berkeley.edu/news/media/releases/2008/08/07_terrorists.shtml.

Blom, Erica, and Raymond de Vries. 2011. Towards local participation in the creation of ethical research guidelines. *Indian Journal of Medical Ethics* VIII: 145–147.

Blyth, Eric. 2004. Patient experiences of an "egg sharing" programme. *Human Fertility* 7 (3): 157–162.

Blyth, Eric, and Lucy Frith. 2009. Donor-conceived people's access to genetic and biographical history: An analysis of provisions in different jurisdictions permitting disclosure of donor identity. *International Journal of Law, Policy, and the Family* 23 (2): 174–191.

Bok, Derek. 2003. *Universities in the Marketplace: The Commercialization of Higher Education*. Princeton University Press.

Bok, Hilary, Kathryn Schill, and Ruth Faden. 2004. Justice, ethnicity, and stem-cell banks. *Lancet* 364: 118–121.

Bolnick, Deborah A., Duana Fullwiley, Troy Duster, Richard S. Cooper, Joan H. Fujimura, Jonathan Kahn, Jay Kaufman, Jonathan Marks, Ann Morning, Alondra Nelson, Pilar Ossorio, Jenny Reardon, Susan M. Reverby, and Kimberly TallBear. 2007. The science and business of genetic ancestry. *Science* 318 (5849): 399–400.

Bolton, Alexander. 2009. Chambliss blocks regulatory pick over animal lawsuits. http://thehill.com/leading-the-news/chambliss-blocks-regulatory-nominee-over-animal-lawsuits-2009-06-28.html.

Bosk, Charles. 2008. *What Would You Do? Juggling Bioethics and Ethnography*. University of Chicago Press.

Boyle, James. 1996. *Shamans, Software, and Spleens: Law and the Construction of the Information Society*. Harvard University Press.

Braun, Kathrin, and Susanne Schultz. 2012. Oöcytes for research: Inspecting the commercialization continuum. *New Genetics & Society* 31 (2): 135–157.

Broder, David. 2001. *Democracy Derailed: Initiative Campaigns and the Power of Money*. Mariner Books.

Browne, Janet. 2003. *Charles Darwin: A Biography*, vol. 2: *The Power of Place*. Princeton University Press.

Browner, Carole, and Carolyn Sargent, eds. 2011. *Reproduction, Globalization, and the State*. Duke University Press.

Brush, Silla, 2005. Hoping to avoid brain drain, states push to finance stem-cell research. *Chronicle of Higher Education* 51 (22): A22.

Brynne, J. A., H. N. Nguyen, and R. A. Reijo Pera. 2009. Enhanced generation of induced pluripotent stem cells from a subpopulation of human fibroblasts. *PLoS ONE* 4 (9): e7118.

Buettinger, Craig. 1997. Women and antivivisection in late nineteenth-century america. *Journal of Social History* 30 (4): 857–872.

Bullard, Robert, and Maxine Waters, eds. 2005. *The Quest for Environmental Justice: Human Rights and the Politics of Pollution.* Sierra Club Books.

Burhansstipanov, L., L. Bernis, J. Kaur, and G. Bernis. 2005. Sample genetic policy language for research conducted with native communities. *Journal of Cancer Education* 20 (Suppl.): 52–57.

Burrell, Jenna. 2009. The field site as a network: A strategy for locating ethnographic research. *Field Methods* 21 (2): 181–199.

Butler, Judith. 2006. *Precarious Life.* Verso.

California Institute for Regenerative Medicine. 2010. *Summary and Recommendations of the CIRM Human iPS Cells Banking Workshop.* San Francisco, California: California Institute for Regenerative Medicine. http://www.cirm.ca.gov/files/PDFs/Publications/iPSC_Banking_Report.pdf.

California Institute for Regenerative Medicine. 2012. *Strategic Plan.* San Francisco, California: California Institute for Regenerative Medicine. http://www.cirm.ca.gov/files/PDFs/Publications/2012_CIRM_stratplan.pdf.

Callaway, Ewen. 2011. Aboriginal genome analysis comes to grips with nature. *Nature* 477: 522–523.

Callon, Michel. 1998a. An essay on framing and overflowing: Economic externalities revisited by sociology. In *The Laws of the Market*, ed. Michel Callon. Blackwell.

Callon, Michel, ed. 1998b. *The Laws of the Market.* Blackwell.

Callon, Michel, and Vololona Rabeharisoa. 2004. Gino's lesson on humanity: Genetics, mutual, entanglements, and the sociologist's role. *Economy and Society* 33 (1): 1–27.

Carroll, Katherine, and Catherine Waldby. 2012. Informed consent and fresh egg donation for stem cell research: Incorporating embodied knowledge into ethical decision-making. *Bioethical Inquiry* 9: 29–39.

Carson, S., D. Eschenbach, G. Lomax, V. Rice, M. Sauer, and R. Taylor. 2010. Proposed oocyte guidelines for stem cell research. *Fertility and Sterility* 94 (7): 2503–2506.

Carter, Joe. 2009. Lament for a bioethics council. http://www.firstthings.com/blogs/firstthoughts/2009/06/18/lament-for-a-bioethics-council/.

Casper, Monica, and Lisa Jean Moore. 2009. *Missing Bodies: The Politics of Visibility.* NYU Press.

Caulfield, Timothy. 2012. Pressured to commercialize: Is the push for science to save the still flailing economy a threat to scientific research? *Scientist,* May 28. http://the-scientist.com/2012/05/28/opinion-pressured-to-commercialize/.

Center for Genetics and Society. 2005. Potential conflicts of interest at the CIRM. Unsigned editorial, April 6, 2005; modified October 3, 2005. http://www.geneticsandsociety.org/article.php?id=318.

Chander, Anupam, and Madhavi Sunder. 2004. The Romance of the Public Domain. *California Law Review* 92: 1331–1374.

Charo, R. Alta. 2005. Realbioethik. *Hastings Center Report* 35 (4): 13–14.

Charo, R. Alta. 2006. Body of research—Ownership and use of human tissue. *New England Journal of Medicine* 355: 1517–1519.

Chen, Haidan. 2009. Stem cell governance in China: From bench to bedside? *New Genetics & Society* 28 (3): 267–282.

Chen, Mel. 2012. *Animacies: Biopolitics, Racial Mattering, and Queer Affect.* Duke University Press.

Cherwin, Arlene. 2007. US: Keeping stem cell research alive. *University World News* 2 (October 21).

Chin, M. H., Mike J. Mason, Wei Xie, Stefano Volinia, et al. (2009). Induced pluripotent stem cells and embryonic stem cells are distinguished by gene expression signatures. *Cell Stem Cell* 5 (1): 111–123.

Cho, Hae-Joang. 1998. Constructing and deconstructing Koreanness. In *Making Majorities: Constituting the Nation in Japan, Korea, China, Malaysia, Fiji, Turkey, and the United States,* ed. Dru Gladney. Stanford University Press.

CIOMS (Council for International Organizations of Medical Sciences). 1985. *International Guiding Principles for Biomedical Research Involving Animals.* http://www.cioms.ch/frame_1985_texts_of_guidelines.htm.

Clarke, Adele, Laura Mamo, Jennifer Ruth Fosket, Jennifer Fishman, and Janet Shim, eds. 2010. *Biomedicalization: Technoscience, Health, and Illness in the U.S.* Duke University Press.

Clarke, Adele, and Virginia Oleson, eds. 1998. *Revisioning Women, Health, and Healing: Feminist, Cultural, and Technoscience Perspectives.* Routledge.

Clough, Patricia Ticineto, and Craig Willse, eds. 2011. *Beyond Biopolitics: Essays on the Governance of Life and Death.* Duke University Press.

Cohen, Joshua. 1996. Procedure and substance in deliberative democracy. In *Democracy and Difference*, ed. Seyla Benhabib. Princeton University Press.

Cohen, Lawrence. 2002. The other kidney: Biopolitics beyond recognition. In *Commodifying Bodies*, ed. Nancy Scheper-Hughes and Loic Wacquant. Sage.

Cohen, Lawrence. 2005. Operability, bioavailability, and exception. In *Global Assemblages*, ed. Aihwa Ong and Stephen Collier. Blackwell.

Congregation for the Doctrine of the Faith. 2008. *Instruction* Dignitas Personae *on Certain Bioethical Questions.* Vatican City.

Cooper, Melinda. 2008. *Life as Surplus: Biotechnology and Capitalism in the Neoliberal Era.* Washington University Press.

Coppin, Dawn. 2002. *Capitalist Pigs: Large-Scale Swine Facilities and the Mutual Construction of Nature and Society.* University of Illinois.

Corless, Peter, Kongsik Kim, Jeon-Suh Park, and Jongahn Kim. 2011. US-Korea Free Trade Agreement and Patent Rights. *Bloomberg Law Reports—Antitrust & Trade* 2 (22): 1–4.

Creager, Angela. 2001. *The Life of a Virus: Tobacco Mosaic Virus as an Experimental Model, 1930–1965.* University of Chicago Press.

Cressey, Daniel. 2009. Top scientist's industry move heralds stem-cell shift. *NATNews* 28 (August). http://www.nature.com/news/2009/090828/full/news.2009.873 .html.

Cressey, Daniel. 2012. Stem cells take root in drug development. *NATNews* 24 (May). http://www.nature.com/news/stem-cells-take-root-in-drug-development-1.10713.

Cussins, Jessica. 2012. Indian surrogate dies amid complications in eighth month of pregnancy. *Biopolitical Times*, May 31. http://www.biopoliticaltimes.org/article .php?id=6243

Cvetkovich, Ann. 2003. *An Archive of Feelings: Trauma, Sexuality, and Lesbian Public Cultures.* Duke University Press.

Cyranoski, David. 2012a. Editor's move sparks backlash. *Nature* 482 (7386): 449–450.

Cyranoski, David. 2012b. Stem-cell therapy takes off in Texas: A boom in unproven procedures is worrying scientists. *Nature* 483: 13–14.

Deloitte Center for Health Solutions. 2008. *Medical Tourism: Consumers in Search of Value.*

Demeritt, David, Sarah Dyer, and James Millington. 2009. PEST or panacea? Science, democracy, and the promise of public participation. *Environment, Politics, and Development Working Paper Series.* Department of Geography, King's College London, Paper #10.

Denton, Paul, and J. Victor Garcia. 2009. Novel humanized murine models for HIV research. *Current HIV/AIDS Reports* 6: 13–19.

Derrida, Jacques. 1996. *Archive Fever: A Freudian Impression.* University of Chicago Press.

De Vries, Raymond, A. Stanczyk, I. F. Wall, R. Uhlmann, L. J. Damschroder, and Scott Yung Ho Kim. 2010. Assessing the quality of democratic deliberation: A case study of public deliberation on the ethics of surrogate consent for research. *Social Science & Medicine* 70 (12): 1896–1903.

Dickenson, Donna. 2005. Human tissue and global ethics. *Genomics, Society, and Policy* 1 (1): 41–53.

Dickenson, Donna. 2006. The lady vanishes: What's missing from the stem cell debate. *Journal of Bioethical Inquiry* 3 (1–2): 43–54.

Dodds, S., and C. Thomson. 2006. Bioethics and democracy: Competing roles of national bioethics organisations. *Bioethics* 20 (3): 326–338.

Dolgin, Elie. 2010. Gap in stem cell funding could drive Australian brain drain. *Nature Medicine* 16 (8): 834.

Dryzek, John. 2005. Deliberative democracy in divided societies. *Political Theory* 33 (2): 218–242.

Duster, Troy. 2006. Explaining differential trust of DNA forensic technology: Grounded assessment or inexplicable paranoia? *Journal of Law, Medicine & Ethics* 34 (2): 293–300.

Ehrich, Kathryn, Clare Williams, and Bobbie Farsides. 2010. Fresh or Frozen? Classifying "spare" embryos for donation to human embryonic stem cell research. *Social Science and Medicine* 71: 2204–2211.

Einhorn, Bruce, Jennifer Veale, and Manjeet Kripalani. 2005. Asia is stem cell central. *Business Week*, January 10.

Einsiedel, Edna, and Hannah Adamson. 2012. Stem cell tourism and future stem cell tourists: Policy and ethical implications. *Developing World Bioethics* 12 (1): 35–44.

Elliott, Carl. 2004. Beyond politics: Why have bioethicists focused on the president's council's dismissals and ignored its remarkable work? *Slate*, March 9. http://www.slate.com/id/2096815/.

Emanuel, Ezekiel. 2005. Undue inducement: Nonsense on stilts? *American Journal of Bioethics* 5 (5): 9–13.

Eng, David. 2010. *The Feeling of Kinship: Queer Liberalism and the Racialization of Intimacy.* Duke University Press.

Engelberg, Alfred, Aaron Kesselheim, and Jerry Avorn. 2009. Balancing innovation, access, and profits—Market exclusivity for biologics. *New England Journal of Medicine* 361: 1917–1919.

Epstein, Steven. 1996. *Impure Science: AIDS, Activism, and the Politics of Knowledge.* University of California Press.

Epstein, Steven. 2007. *Inclusion: The Politics of Difference in Medical Research.* University of Chicago Press.

ESCAP (Economic and Social Commission for Asia and the Pacific). 2011. *Strengthening the Governance of National Innovation Systems.*

Ethics Committee of the American Society for Reproductive Medicine. 2000. Financial incentives in recruitment of oocyte donors. *Fertility and Sterility* 74 (2): 216–220.

Ethics Committee of the American Society for Reproductive Medicine. 2003. Family members as gamete donors and surrogates. *Fertility and Sterility* 20 (5): 1124–1130.

Ethics Committee of the American Society for Reproductive Medicine. 2007. Financial compensation of oocyte donors. *Fertility and Sterility* 88 (2): 305–309.

Evans, Martin. 2001. The cultural mouse. *Nature Medicine* 7: 1081–1083.

Evans, M. J., and M. H. Kaufman. 1981. Establishment in culture of pluripotential cells from mouse embryos. *Nature* 292: 154–156.

Everts, Sarah. 2007. Reducing animal testing using stem cells. [Online Issue.] *Chemical and Engineering News* 85 (3): 15.

Executive Office of the President of the United States. 2007. *Advancing Tissue Science and Engineering: A Foundation for the Future. A Multi-Agency Strategic Plan.* http://www.tissueengineering.gov/welcome-s.php.

Faiola, Anthony. 2006. Koreans "blinded" to truth about claims on stem cells. *Washington Post Foreign Service*, January 13, A10.

Feldman, Ilana, and Miriam Tiktin, eds. 2010. *In the Name of Humanity: The Government of Threat and Care.* Duke University Press.

Feller, David Allan. 2009. Dog fight: Darwin as animal advocate in the antivivisection controversy of 1875. *Studies in History and Philosophy of Science, Part C: Studies in History and Philosophy of Biological and Biomedical Sciences* 40 (4): 265–271.

Flanagan, Jerry. 2005. Letter from the Foundation for Taxpayer and Consumer Rights to Robert Klein, Chair, Independent Citizen's Oversight Committee, March 1, 2005.

Fletcher, Matthew. 2010. Havasupai Press Release on the Arizona State DNA Settlement. *Turtle Talk.* http://turtletalk.wordpress.com/.

Fortun, Michael. 2008. *Promising Genomics: Iceland and decode Genetics in a World of Speculation.* University of California Press.

Foucault, Michel. 1998. *The History of Sexuality*, volume 1. Penguin.

Fourcade, Marion. 2009. *Economists and Societies: Discipline and Profession in the United States, Britain, and France, 1890s to 1990s.* Princeton University Press.

Fox, R. C., J. P. Swazey, and J. C. Watkins. 2008. *Observing Bioethics.* Oxford University Press.

Franklin, R., and R. G. Gosling. 1953. Molecular configuration in sodium thymonucleate. *Nature* 171: 740–741.

Franklin, Sarah. 1997. *Embodied Progress: A Cultural Account of Assisted Reproduction.* Routledge.

Franklin, Sarah. 2005. Stem Cells R Us: Emergent life forms and the global biological. In *Global Assemblages: Technology, Politics, and Ethics as Anthropological Problems*, ed. A. Ong and S. J. Collier. Blackwell.

Franklin, Sarah. 2006. The cyborg embryo: Our path to transbiology. *Theory, Culture & Society* 23 (7–8): 167–187.

Franklin, Sarah. 2007. *Dolly Mixtures: The Remaking of Genealogy*. Duke University Press.

Franklin, Sarah. 2011. Not a flat world: The future of cross-border reproductive care. *Reproductive Biomedicine Online* 23: 814–816.

Franklin, Sarah, and Margaret Lock. 2003. *Remaking Life and Death: Toward an Anthropology of the Biosciences*. School of American Research.

Franklin, Sarah, and Celia Roberts. 2006. *Born and Made: An Ethnography of Pre-implantation Genetic Diagnosis*. Princeton University Press.

Furcht, Leo, and William Hoffman. 2008. *The Stem Cell Dilemma: Beacons of Hope or Harbingers of Doom?* Arcade.

Gabriel, Monica. 2009. GE Plans to Use Human Embryonic Stem Cells, Not Lab Rats, to Test Drug Toxicity. http://ww.cnsnews.com/public/content/article.aspx?RsrcID-5074.

Galef, Julia. 2011. You say embryo, I say parthenote: Stem cells from unfertilized eggs may be too tightly regulated. *Scientific American* (November). http://www.scientificamerican.com/article.cfm?id=you-say-embryo-i-say-parthenote.

Galison, Peter. 1994. The ontology of the enemy: Norbert Wiener and the cybernetic vision. *Critical Inquiry* 21 (1): 228–266.

Galison, Peter. 1997. *Image and Logic: A Material Culture of Microphysics*. University of Chicago Press.

Galison, Peter. 2004a. *Einstein's Clocks, Poincaré's Maps: Empires of Time*. Norton.

Galison, Peter. 2004b. Removing knowledge. *Critical Inquiry* 3: 229–243.

Ganchoff, Chris. 2008. Speaking for stem cells: Biomedical activism and emerging forms of patienthood. *Advances in Medical Sociology* 10: 225–245.

Garilov, Svetlana, Darja Maroit, Nataki C. Douglas, Robert W. Prosser, Imran Khalid, Mark V. Sauer, Donald W. Landry, Gordana Vunjak-Novakovic, and Virginia

E. Papaioannou. 2011. Derivation of two new human embryonic stem cell lines from nonviable human embryos. *Stem Cell International.* http://www.hindawi.com/journals/sci/2011/765378/.

Garvey, Megan. 2005. California's stem cell bid stuck in neutral. *Los Angeles Times,* May 23.

Geransar, Rose, and Edna Einsiedel. 2008. Evaluating online direct-to-consumer marketing of genetic tests: Informed choices or buyers beware? *Genetic Testing* 12 (1): 13–23.

Geron Corporation. 2009a. GE Healthcare and Geron announce exclusive global agreement to commercialize stem cell drug discovery technologies. http://www.geron.com/media/presview.aspx?id=1181.

Geron Corporation. 2009b. Geron received FDA clearance to begin world's first human clinical trial of embryonic stem cell-based therapy. Press release: http://www.geron.com/media/presview.aspx?id=1148.

Gibbons, Michael, 1999. Science's new social contract with society. *Nature* 402: C81–84.

Gilboy, N., P. Tanabe, D. A. Travers, A. M. Rosenau, and D. R. Eitel. 2005. *Emergency Severity Index, Version 4: Implementation Handbook.* AHRQ Publication No. 05-0046-2, May 2005. Agency for Healthcare Research and Quality, Rockville, MD. http://www.ahrq.gov/research/esi/.

Ginsburg, Faye, and Rayna Rapp, eds. 1995. *Conceiving the New World Order: The Global Politics of Reproduction.* University of California Press.

Gitter, Donna. 2004. Ownership of human tissue: A proposal for federal recognition of human research participants' property rights in their biological material. *Washington and Lee Law Review* 61 (1): 257–345.

Glave, Dianne, and Mark Stoll, eds. 2005. *To Love the Wind and the Rain: African Americans and Environmental History.* University of Pittsburgh Press.

Glenn, Evelyn Nakano. 2004. *Unequal Freedom: How Race and Gender Shaped American Citizenship and Labor.* Harvard University Press.

Glenn, Evelyn Nakano. 2010. *Forced to Care: Coercion and Caregiving in America.* Harvard University Press.

Goldberger, Paul. 2011. The Skyline: "Laboratory Conditions." *New Yorker* (September 19): 88–89.

Goodwin, Michele, 2011. Reproductive carrots and sticks. *Scholar and Feminist Online* 9.1–9.2, fall 2010–spring 2011.

Goozner, Merrill, 2006. Innovation in biomedicine: Can stem cell research lead the way to affordability? *PLoS Medicine* 3 (5): e126.

Gordin, Michael. 1992. Historical origins of the Nuremberg Code. In *The Nazi Doctors and the Nuremberg Code: Human Rights in Human Experimentation*, ed. George Annas and Michael Grodin. Oxford University Press.

Gottweis, Herbert, Brian Salter, and Catherine Waldby, eds. 2009. *The Global Politics of Human Embryonic Stem Cell Science: Regenerative Medicine in Transition*. Palgrave.

Grady, Christine. 2001. Money for research participation: Does it jeopardize informed consent? *American Journal of Bioethics* 1 (2): 40–44.

Greek, C. Ray, and Jean Swingle Greek. 2003. *Specious Science: How Genetics and Evolution Reveal Why Medical Research on Animals Harms Humans*. Continuum.

Greely, Henry, Daniel Riordan, Nanibaa' Garrison, and Joanna Mountain. 2006. Family ties: The use of DNA offender databases to catch offenders' kin. *Journal of Law, Medicine & Ethics* 34 (2): 248–262.

Grimm, Daniel. 2007. The demographics of genetic surveillance: Familial DNA testing and the hispanic community. *Columbia Law Review* 107: 1164–1194.

Gutmann, Amy, and Dennis Thompson. 2004. *Why Deliberative Democracy?* Princeton University Press.

Habermas, Jürgen. 1996. *Between Facts and Norms: Contributions to a Discourse of Theory of Law and Democracy*. MIT Press.

Haimes, Erica. 2002. What can the social sciences contribute to the study of ethics? Theoretical, empirical, and substantive considerations. *Bioethics* 16: 89–113.

Haimes, Erica, Ken Taylor, and Ilke Turkmendag, 2012. Eggs, ethics, and exploitation? Investigating women's experiences of an "egg sharing" scheme. *Sociology of Health and Illness* 34 (8).

Halberstam, Judith. 2011. *The Queer Art of Failure*. Duke University Press.

Hall, Zach. 2008. Stem cells and Leonardo's cave. *Cell Stem Cell* 2: 536–537.

Hamilton, David. 2007. Stem-cell "brain drain" or "brain gain"? *VentureBeat News*, August 17. http://venturebeat.com/2007/08/17/stem-cell-brain-drain-or-brain-gain/.

Hammonds, Evelynn M., and Rebecca Herzig, eds. 2008. *The Nature of Difference: Sciences of Race in the United States from Jefferson to Genomics*. MIT Press.

Haraway, Donna. 1989. *Primate Visions: Gender, Race, and Nature in the World of Modern Science*. Routledge, Chapman, & Hall.

Haraway, Donna. 1990a. *Simians, Cyborgs, and Women: The Reinvention of Nature*. Routledge.

Haraway, Donna. 1990b. Situated knowledges: The science question in feminism and the privilege of partial perspective. In *Simians, Cyborgs, and Women: The Reinvention of Nature*. Routledge.

Haraway, Donna. 1997. *Modest_Witness@Second_Millennium: FemaleMan®_Meets_Oncomouse*. Routledge.

Haraway, Donna. 2007. *When Species Meet*. University of Minnesota Press.

Harmon, Katherine. 2009. Induced pluripotent stem cells created from fat cells. *Scientific American Online*, September 8. http://www.scientificamerican.com/article.cfm?id=stem-cells-from-fat-cells

Harris, Cheryl. 1993. Whiteness as property. *Harvard Law Review* 106 (8): 1707–1791.

Harris, D. T., and I. Rogers. 2007. Umbilical cord blood: A unique source of pluripotent stem cells for regenerative medicine. *Current Stem Cell Research & Therapy* 2 (4): 301–309.

Harrison, Charlotte. 2002. Neither Moore nor the market: Alternative models for compensating contributors of human tissue. *American Journal of Law & Medicine* 28: 77–105.

Hayden, Cori. 2007. Taking as giving: Bioscience, exchange, and the politics of benefit-sharing. *Social Studies of Science* 37: 729–758.

Hayden, Erika Check. 2012. Informed consent: A broken contract. *Nature* 486: 312–314.

Hayden, Erika Check. 2008. Stem cells: The 3-billion-dollar question. *Nature* 453: 18–21.

Henderson, Gail, Larry Churchhill, Arlene Davis, Michele Easter, Christine Grady, Steven Joffe, Nancy Kass, et al. 2007. Clinical trials and medical care: Defining the therapeutic misconception. *PLoS Medicine* 4 (11): 1735–1738.

Hendriks, Frank. 2010. *Vital Democracy: A Theory of Democracy in Action.* Oxford University Press.

Herzig, Rebecca. 2005. *Suffering for Science: Reason and Sacrifice in Modern America.* Rutgers University Press.

Highfield, Roger. 2001. Brain drain to Britain for embryo research. *Telegraph,* July 18.

Hiltzik, Michael. 2011. California stem cell agency needs to study itself. *Los Angeles Times,* December 7.

Hird, Myra. 2009. *The Origins of Sociable Life: Evolution after Science Studies.* Palgrave Macmillan.

Ho, W. Calvin, Benjamin Capps, and Teck Chuan Voo. 2010. Stem cell science and its public: The case of Singapore. *East Asian Science, Technology, and Society: An International Journal* 4 (1): 7–29.

Hochschild, Arlie. 2003. *The Managed Heart: Commercialization of Human Feeling, Twentieth Anniversary Edition.* University of California Press.

Holden, Constance. 2005. U.S. States offer Asia stiff competition. *Science* 307 (5710): 662–663.

Holden, Kerry, and David Demeritt. 2008. Democratising science? The politics of promoting biomedicine in Singapore's developmental state. *Environment and Planning, Series D: Society & Space* 26 (1): 68–86.

Holland, Suzanne, Karen Lebacqz, and Laurie Zoloth, eds. 2001. *The Human Embryonic Stem Cell Debate.* MIT Press.

Holloway, Karla. 2011. *Private Bodies, Public Texts: Race, Gender, and a Cultural Bioethics.* Duke University Press.

Holm, Søren. 2006. Who should control the use of human embryonic stem cell lines: A defense of the donors' ability to control. *Journal of Bioethical Inquiry* 3 (1–2): 55–68.

Hong, Sungook. 2008. The Hwang scandal that "shook the world of science." *East Asian Science, Technology, and Society* 2: 1–7.

Høyer, Klaus. 2009. Informed consent: The making of a ubiquitous rule in medical practice. *Organization* 16 (2): 267–288.

Høyer, Klaus. 2004. Ambiguous gifts: Public anxiety, informed consent, and biobanks. In *Genetic Databases: Socio-Ethical Issues in the Collection and Use of DNA*, ed. R. Tutton and O. Corrigan. Routledge.

Huangfu, Danwei, René Maehr, Wenjun Guo, Astrid Eijkelenboom, Melinda Snitow, Alice Chen, and Douglas Melton. 2008. Induction of pluripotent stem cells by defined factors is greatly improved by small molecule compounds. *Nature Biotechnology* 26 (7): 794–797.

Hurlbut, William B. 2005. Altered nuclear transfer as a morally acceptable means for the procurement of human embryonic stem cells. *Perspectives in Biology and Medicine* 48 (2): 211–228.

Hussein, Ghaiath M. A. 2009. Democracy: The forgotten challenge for bioethics in the developing countries. *BMC Medical Ethics* 10: 3.

Hwang, Woo-Suk, et al. 2004. Evidence of a pluripotent human embryonic stem cell line derived from a cloned blastocyst. *Science* 303: 1669–1674.

Hwang, Woo-Suk, et al. 2005. Patient-specific embryonic stem cells derived from human SCNT blastocysts. *Science* 308: 1777–1783.

Hyun, Insoo. 2008. Stem cells from skin cells: The ethical questions. *Hastings Center Report* 38 (1): 20–22.

Ikemoto, Lisa. 2009. Eggs as capital: Human egg procurement in the fertility industry and the stem cell research enterprise. *Signs: Journal of Women in Culture and Society* 34: 763–781.

Irving, Michelle, Allison Tong, Stephen Jan, Alan Cass, John Rose, Steven Chadban, Richard Allen Jnathan Craig, Germaine Wong, and Kirsten Howard, 2011. Factors that influence the decision to be an organ donor: A systematic review of the qualitative literature. *Nephrology Dialysis Transplantation Online*, December 21. doi: 10.1093/ndt/gfr683.

Irwin, Alan. 2001. Constructing the scientific citizen: Science and democracy in the biosciences. *Public Understanding of Science* 10 (1): 1–18.

Irwin, Alan. 2006. The politics of talk: Coming to terms with the "new" scientific governance. *Social Studies of Science* 36 (2): 299–320.

ISSCR (International Society for Stem Cell Research). 2008. *Guidelines for the Clinical Translation of Stem Cells*. ISSCR Press. http://www.isscr.org.

Ito, Ryoji, Takeshi Takahashi, Ikumi Katano, and Mamoru Ito. 2012. Current advances in humanized mouse models. *Cellular & Molecular Immunology* 9: 208–214.

Ivison, Duncan. 2002. *Postcolonial Liberalism*. Cambridge University Press.

Jackson, Zakkiyah. 2011. Who cuts the border? Race and the future of animal studies. Paper presented at Why the Animal? Queer Animalities, Indigenous Natu-recultures, and Critical Race Approaches to Animal Studies, Center for Science, Technology, and Medicine in Society, UC Berkeley, April 2011.

Jain, Sarah Lochlann. 2006. Living in prognosis. *Representations* 98 (1): 77–92.

Jasanoff, Sheila. 2002. The life sciences and the rule of law. *Journal of Molecular Biology* 319: 891–899.

Jasanoff, Sheila. 2003. Technologies of humility: Citizen participation in governing science. *Minerva* 41 (3): 223–244.

Jasanoff, Sheila, ed. 2004. *States of Knowledge: The Co-Production of Science and Social Order*. Routledge.

Jasanoff, Sheila. 2005. *Designs on Nature: Science and Democracy in Europe and the United States*. Princeton University Press.

Jensen, David. 2005–2012. *California Stem Cell Report*. http://californiastemcellreport.blogspot.com/

Jesudason, Sujatha, 2011. The latest case of reproductive carrots and sticks: Race, abortion, and sex selection. *Scholar and Feminist Online* 9.1–9.2, fall 2010–spring 2011.

Johnson, Judith, and Erin Williams, 2006. Stem cell research: State initiatives. *Congressional Research Services Report for Congress* RL 33524.

Johnson, Simon. 2009. The quiet coup. *Atlantic* (May).

Jones, David. 2009. What does the British public think about human–animal hybrid embryos? *Journal of Medical Ethics* 35: 168–170.

Jonsen, Albert, and Stephen Toulmin. 1988. *The Abuse of Casuistry: A History of Moral Reasoning*. University of California Press.

Jung, Kyu Won. 2010. Regulation of human stem cell research in South Korea. *Stem Cell Reviews and Reports* 6 (3): 340–344.

Kahn, Jeffrey. 2001. Stem Cells and a New Brain Drain. *CNN Health*. http://articles.cnn.com/2001-07-23/health/ethics.matters_1_cell-research-funding-for-such-research-human-embryos?_s=PM: HEALTH.

Kao, John. 2009. Tapping the world's innovation hot spots. *Harvard Business Review* (March): 1–7.

Kaplan, Caren. 2006. Precision targets: GPS and the militarization of U.S. consumer identity. *American Quarterly* 58 (3): 693–714.

Kaplan, Caren, Norma Alarcón, and Minoo Moallem, eds. 1999. *Between Women and Nation: Nationalisms, Transnational Feminisms, and the State*. Duke University Press.

Kaufman, Sharon. 2006. *And a Time to Die: How American Hospitals Shape the End of Life*. University of Chicago Press.

Kennedy, Donald. 2006. Editorial retraction. *Science* 311 (5759): 335.

Kennedy, Michael. 2011. Arab Spring, Occupy Wall Street, and historical frames: 2011, 1989, and 1968. *Jadaliyya*, October 11. http://www.jadaliyya.com/pages/index/2853/arab-spring-occupy-wall-street-and-historical-fram.

Kelly, Christine. 2011. Making "care" accessible: Personal assistance for disabled people and the politics of language. *Critical Social Policy* 31 (4): 562–582.

Kenney, N. J., and M. L. McGowan. 2010. Looking back: Egg donors' retrospective evaluations of their motivations, expectations, and experiences during their first donation cycle. *Fertility and Sterility* 93 (2): 455–466.

Khushf, G. 2007. Upstream ethics in nanomedicine: A call for research. *Nanomedicine; Nanotechnology, Biology, and Medicine* 2 (4): 511–521.

Kian, Catherine, and Tien Sim Leng. 2005. The Singapore approach to human stem cell research, therapeutic and reproductive cloning. *Bioethics* 19 (3): 290–303.

Kim, Mi-Kyung. 2009. Oversight framework over oocyte procurement for somatic cell nuclear transfer: comparative analysis of the Hwang Woo Suk case under South Korean bioethics law and U.S. guidelines for human embryonic stem cell research. *Theoretical Medicine and Bioethics* 30: 367–384.

Kirksey, Eben, and Stefan Helmreich. 2010. The emergence of multispecies ethnography. *Cultural Anthropology* 25 (4): 545–576.

Kittay, Eva Feder, and Licia Carson eds. 2010. *Cognitive Disability and Its Challenge to Moral Philosophy.* Wiley-Blackwell.

Klimanskaya, I., Y. Chung, S. Becker, S. J. Lu, and R. Lanza. (2006). Human embryonic stem cell lines derived from single blastomeres. *Nature* 444 (7118): 481–485.

Klitzman, R., and M. V. Sauer. 2009. Payment of egg donors in stem cell research in the USA. *Reproductive Biomedicine Online* 18 (5): 603–608.

Knorr Cetina, Karin. 1999. *Epistemic Cultures: How the Sciences Make Knowledge.* Harvard University Press.

Kohler, Robert. 1994. *Lords of the Fly: Drosophila Genetics and the Experimental Life.* University of Chicago Press.

Kohn, Eduardo. 2007. How dogs dream: Amazonian natures and the politics of transspecies engagement. *American Ethnologist* 34 (1): 3–24.

Konrad, Monica. 2005. *Nameless Relations: Anonymity, Melanesia and Reproductive Gift Exchange among British Ova Donors and Recipients.* Berghahn.

Korobkin, Russell. 2006. Embryonic histrionics: A critical evaluation of the Bush stem cell funding policy and the congressional alternative. *Jurimetrics* 47: 1–29.

Kotz, Joanne. 2012. Bringing patient data into the open. *Science-Business Exchange* 5 (25). http://www.nature.com/scibx/journal/v5/n25/full/scibx.2012.644.html

Krones, T., T. Samusch, S. Weber, I. Budiner, A. Busch, F. Knappertsbusch, E. Schluter, and C. Hauskeller. 2008. Brain drain in stem cell research? The views and attitudes of stem cell researchers in Germany. *Bundesgesundheitsblatt–Gesendheitsforschung–Gesundheitsschutz* 51 (9): 1039–1049.

LaDuke, Winona. 1999. *All Our Relations: Native Struggles for Land and Life.* South End.

LaFollette, Marcel. 2000. The evolution of the "scientific misconduct" issue: An historical overview. *Experimental Biology and Medicine* 224 (4): 211–215.

Landecker, Hannah. 2007. *Culturing Life: How Cells Became Technologies.* Harvard University Press.

Langston, James. 2011. CIRM's consolidated IP policy: Will it promote collaboration? *Santa Clara Law Review* 51 (2): 671–704.

Latour, Bruno. 1988. *The Pasteurization of France*. Harvard University Press.

Latour, Bruno. 2004. Why has critique run out of steam? From matters of fact to matters of concern. *Critical Inquiry* 30 (winter): 225–248.

Ledford, Heidi. 2009. Hybrid embryos fail to live up to stem-cell hopes. *NATNews* 3 (February). http://www.nature.com/news/2009/090203/full/457642b.html.

Lee, Cheol-Sung, and Andrew Schrank. 2010. incubating innovation or cultivating corruption? The developmental state and the life sciences in Asia. *Social Forces* 88 (3): 1231–1256.

Lee, Peter. 2008. Contracting to preserve open science: Consideration-based regulation in patent law. *Emory Law Journal* 58: 889–975.

Lee, Wei Ling, 2006. What ails biomedical research in Singapore? *Straits Times*, November 4.

Leem, So Yeon, and Jin Hee Park. 2008. Rethinking women and their bodies in the age of biotechnology: Feminist commentaries on the Hwang affair. *East Asian Science, Technology and Society* 2: 9–26.

Lemke, Amy, Joel Wu, Carol Waudby, Jill Pulley, Carol Somkin, and Susan Trinidad. 2010/11. Community engagement in biobanking: Experiences from the Emerge Network. *Genomics, Society, and Policy* 6 (3): 50–67.

Lemke, Thomas. 2011. *Biopolitics: An Advanced Introduction*. NYU Press.

Lengerke, Claudia, and George Daley. 2010. Autologous blood cell therapies from pluripotent stem cells. *Blood Reviews* 24 (1): 27–37.

Lesch, John. 2007. *The First Miracle Drugs: How the Sulfa Drugs Transformed Medicine*. University of Oxford Press.

Lie, John. 2004. *Modern Peoplehood*. Harvard University Press.

Lim, Sylvia, and Ho, Calvin, 2003. The ethical position of Singapore on embryonic stem cell research. *SMA News* 35 (6).

Lister, Ryan, Mattia Pelizzola, Yasuyuki Kida, R. David Hawkins, Josephy Nery, Gary Hon, Jessica Antosiewicz-Bourget, et al. 2011. Hotspots of aberrant epigenomic reprogramming in human induced pluripotent stem cells. *Nature* 471: 68–73.

Little Hoover Commission. 2009. *Stem Cell Research: Strengthening Governance to Further the Voters' Mandate*. Sacramento, California.

Liu, Jennifer. 2008. Asia modern: Stem cells, ethics, and contemporary Taiwan. PhD dissertation, UC Berkeley-UCSF Medical Anthropology Program.

Lock, Margaret. 2001. *Twice Dead: Organ Transplants and the Reinvention of Death*. University of California Press.

Lock, Margaret, and Vinh-Kim Nguyen. 2010. *An Anthropology of Biomedicine*. Blackwell.

Lock, Margaret, Alan Young, and Alberto Cambrosio, eds. 2000. *Living and Working with the New Medical Technologies: Intersections of Inquiry*. Cambridge University Press.

Loew, Franklin, and Bennett Cohen. 2002. Laboratory animal medicine: Historical perspectives. In *Laboratory Animal Medicine*, second edition, ed. James Fox, Lynn Anderson, Franklin Loew, and Fred Quimby. Academic.

Lowry, Ilana. 2000. Trustworthy knowledge and desperate patients: Clinical tests for new drugs from cancer to AIDS. In *Living and Working with the New Medical Technologies: Intersections of Inquiry*, ed. Margaret Lock, Alan Young, and Alberto Cambrosio. Cambridge University Press

Luker, Kristin. 1984. *Abortion and the Politics of Motherhood*. University of California Press.

Lunshof, Jeantine, Ruth Chadwick, Daniel Vorhaus, and George Church. 2008. From genetic privacy to open consent. *Nature Reviews: Genetics* 9: 406–411.

Lynch, Michael. 1999. Archives in formation: Privileged spaces, popular archives and paper trails. *History of the Human Sciences* 12 (2): 65–87.

Magnus, David. 2012. David Magnus corrects the record regarding AJOB. Editor's blog. *American Journal of Bioethics* (February 23). http://www.bioethics.net/2012/02/david-magnus-corrects-the-record-regarding-ajob/.

Magnus, David, and Mildred Cho. 2005. Issues in oocyte donation for stem cell research. *Science* 308 (5729): 1747–1748.

Maienschein, Jane. 2007. Untangling debates about science and religion. In *The Panda's Black Box: Opening Up the Intelligent Design Controversy*, ed. Nathanial Comfort. Johns Hopkins University Press.

Mali, Prashant, and Linzhao Cheng. 2011. Human cell engineering: Cellular reprogramming and genome editing. *Stem Cells* 30 (1): 75–81.

Martin, Brian. 1992. Scientific fraud and the power structure of science. *Prometheus* 10 (1): 83–98.

Martin, Gail. 1981. Isolation of a pluripotent cell line from early mouse embryos cultured in medium conditioned by teratocarcinoma stem cells. *Proceedings of the National Academy of Sciences of the United States of America* 78 (12): 7634–7638.

Martin, Mark. 2002. Davis OKs stem cell research: California is first state to encourage studies. *San Francisco Chronicle*, September 23, A1.

Mason, Mary Ann, Marc Goulden, and Karie Frasch. 2011. Keeping women in the science pipeline. *Annals of the American Academy of Political and Social Science* 638 (1): 141–162.

Mathews, Debra, Gregory Graff, Krishanu Saha, and David Winickoff. 2011. Access to stem cells and data: Persons, property rights, and scientific progress. *Science* 331 (6018): 725–727.

Mbembe, Achille. 2001. *On the Postcolony*. University of California Press.

Mbembe, Achille. 2003. Necropolitics. *Public Culture* 15 (1): 11–40.

McGee, Glenn. 2006. Editorial retraction. *American Journal of Bioethics* 6 (1): W33.

McKenna, Michael. 2009. Compatibilism. In *The Stanford Encyclopedia of Philosophy* (winter 2009 edition), ed. Edward Zalta. http://plato.stanford.edu/archives/win2009/entries/compatibilism/.

Mello, Michelle, and Leslie Wolf. 2010. The Havasupai Indian tribe case—Lessons for research involving stored biologic samples. *New England Journal of Medicine* 363: 204–207.

Merz, Jon, David Magnus, Mildred Cho, and Arthur Caplan. 2002. Protecting subjects' interests in genetics research. *American Journal of Human Genetics* 70 (4): 965–971.

Meslin, E. M., and M. K. Cho. 2010. Research ethics in the era of personalized Medicine: Updating science's contract with society. *Public Health Genomics* 13 (6): 378–384.

Meyer, Jean-Baptiste, David Kaplan, and Jorge Charum. 2001. Scientific nomadism and the new geopolitics of knowledge. *International Social Science Journal* 53 (168): 309–321.

Miki, Toshio, and Stephen Strom. 2006. Amnion-derived pluripotent/multipotent stem cells. *Stem Cell Reviews* 2: 133–142.

Milde, Margo, and John Yates. 2009. Noose set to tighten on Los Angeles Purebred Dogs: AKC disqualification shows entire nation the danger of compromise, apathy or agreeing to biased task forces. *American Sporting Dog Alliance.* http://www.americansportingdogalliance.org

Miller, Alice. 2000. Sexual but not reproductive: Exploring the junction and disjunction of sexual and reproductive rights. *Health and Human Rights* 4 (2): 68–109.

Mirowski, Philip. 2011. *Science-Mart: Privatizing American Science.* Harvard University Press.

Mol, Annemarie. 2002. *The Body Multiple: Ontology in Medical Practice.* Duke University Press.

Mol, Annemarie. 2008. *The Logic of Care: Health and the Problem of Patient Choice.* Routledge.

Mol, Annemarie, Ingunn Moser, and Jeannette Pols. 2010. Care: Putting practice into theory. In *Care in Practice: On Tinkering in Clinics, Homes, and Farms,* ed. Annemarie Mol, Ingunn Moser, and Jeannette Pols. Transcript.

Moreno, Jonathan. 2006. *Mind Wars: Brain Research and National Defense.* Dana.

Moreno, Jonathan. 2011. *The Body Politic: The Battle over Science in America.* Bellevue.

Moreno, Jonathan, and Sam Berger, eds. 2010. *Progress in Bioethics: Science, Policy, and Politics.* MIT Press.

Morrison, Michael. 2011. Comment: The patenting of human embryonic stem cells. *Genomics Network,* November 27. http://www.genomicsnetwork.ac.uk/egenis/news/comment/title,25326,en.html.

Motluk, Alison. 2011. Canadian court bans anonymous sperm and egg donation. *Nature News.* Published online May 27. http://www.nature.com/news/2011/110527/full/news.2011.329.html.

Mouffe, Chantal. 1999. Deliberative democracy or agonistic pluralism? *Social Research* 66 (3): 745–758.

Mowery, David, and Bhaven Sampat. 2004. The Bayh-Dole Act of 1980 and university–industry technology transfer: A model for other OECD governments? *Journal of Technology Transfer* 30 (1): 115–127.

National Bioethics Advisory Commission. 1999. *Ethical Issues in Human Stem Cell Research: Executive Summary.*

National Research Council. 2005. *Guidelines for Human Embryonic Stem Cell Research.* National Academies Press.

National Research Council. 2011. *Toward Precision Medicine: Building a Knowledge Network for Biomedical Research and a New Taxonomy of Disease.* National Academies Press.

Nelson, Alondra. 2011. *Body and Soul: The Black Panther Party and the Fight against Medical Discrimination.* University of Minnesota Press.

Nguyen, Vinh-Kim. 2010. *The Republic of Therapy: Triage and Sovereignty in West Africa's Time of AIDS.* Duke University Press.

Nickerson, Colin. 2009. More accuracy seen in living cell stand-ins for human organs. *Boston Globe*, March 30.

Noggle, Scott, Ho-Lim Fung, Athurva Gore, Hector Martinez, et al. 2011. Human oocytes reprogram somatic cells to a pluripotent state. *Nature* 478: 70–75.

Norsigian, Judy. 2005. Egg donation for IVF and stem cell research: Time to weigh the risks to women's health. *Different Takes* 33. http://www.etopiamedia.net/empnn/pdfs/norsigian1.pdf.

Not Dead Yet. 1999. Disability activists protest singer appointment. http://www.dimenet.com/actions/archive.php?mode=N&id=51.

Nowotny, H., P. Scott, and M. Gibbons. 2001. *Re-thinking Science: Knowledge and the Public in an Age of Uncertainty.* Polity.

Nuffield Council on Bioethics. 2011. *Human Bodies: Donation for Medicine and Research.*

Nuremberg Military Tribunal. 1949. Nuremberg Code: Directives for human experimentation. In *Trials of War Criminals before the Nuremberg Military Tribunals Under Control Council Law*, volume 2. US Government Printing Office.

Oakley, Barbara, Ariel Knafo, Guruprasad Madhavan, and David Wilson, eds. 2012. *Pathological Altruism*. Oxford University Press.

OECD (Organization for Economic Co-operation and Development). 2012. *Main Science and Technology Indicators*, volume 2011/12.

Office of Laboratory Animal Welfare. 2002. Public health service policy on humane care and use of laboratory animals. http://grants.nih.gov/grants/olaw/olaw.htm.

Oliver, Michael. 1990. *The Politics of Disablement: A Sociological Approach.* Macmillan.

Ong, Aihwa. 2005. Ecologies of expertise: Assembling flows, managing citizenship. In *Global Assemblages: Technology, Politics and Ethics as Anthropological Problems*, ed. A. Ong and S. J. Collier. Blackwell.

Ong, Aihwa. 2007. *Please stay*: Pied-a-terre subjects in the megacity. *Citizenship Studies* 11 (1): 83–93.

Ossorio, Pilar, and Troy Duster. 2005. Race and genetics: Controversies in biomedical, behavioral, and forensic sciences. *American Psychologist* 60 (1): 115–128.

Overland, Martha Ann. 2007. British stem-cell scientist is latest prominent researcher to leave Singapore. *Chronicle of Higher Education* (December 13).

Owen-Smith, Jason, and Jennifer McCormick. 2006. Correspondence: An international gap in human ES research. *Nature Biology* 24 (4): 391–392.

Parayil, Govindan. 2005. From "Silicon Island" to "Biopolis of Asia": Innovation policy and shifting competitive strategy in Singapore. *California Management Review* 47 (2): 50–73.

Parens, Erik, and Adrienne Asch, eds. 2000. *Prenatal Testing and Disability Rights.* Georgetown University Press.

Park, Han Woo, Heung Deug Hong, and Loet Leydesdorff. 2005. A comparison of the knowledge-based innovation systems in the economies of South Korea and the Netherlands using triple helix indicators. *Scientometrics* 65 (1): 3–27.

Park, In-Hyun, Natasha Arora, Hongguang Huo, Nimet Maherali, Tim Ahfedt, Akiko Shimamura, M. William Lensch, Chad Cowan, Konrad Hochedlinger, and George Daley. 2008. Disease-specific induced pluripotent stem (iPS) cells. *Cell* 135 (5): 877–886.

Park, Katharine. 2006. *Secrets of Women: Gender, Generation, and the Origins of Human Dissection.* Zone.

Parry, Bronwyn. 2005. From the corporeal to the informational: Exploring the scope of benefit sharing agreements and their applicability to sequence databases. *Wissenschaftsethik und Technikfolgenbeurteilung* 24: 73–91.

Parthasarathy, Shobita. 2010. Breaking the expertise barrier: Understanding activist strategies in science and technology policy domains. *Science & Public Policy* 37 (5): 355–367.

Penner, James. 2010. *Pinks, Pansies, and Punks: The Rhetoric of Masculinity in American Literary Culture.* Indiana University Press.

Perez, Laura. 2007. *Chicana Art: The Politics of Spiritual and Aesthetic Alterities.* Duke University Press.

Pérez-Cano, R., J. Vranckx, J. Lasso, C. Calabrese, B. Merck, A. Milstein, E. Sassoon, E. Delay, and E. Weiler-Mithoff. 2012. Prospective trial of adipose-derived regenerative cell (ADRC)-enriched fat grafting for partial mastectomy defects: The RESTORE-2 trial. *European Journal of Surgical Oncology* 38 (5): 382–389.

Peters, Ted. 2007. *The Stem Cell Debate.* Fortress.

Petryna, Adriana. 2002. *Life Exposed: Biological Citizens after Chernobyl.* Princeton University Press.

Petryna, Adriana. 2009. *When Experiments Travel: Clinical Trials and the Global Search for Human Subjects.* Princeton University Press.

Pickering, Susan, Stephen Minger, Minal Patel, Hannah Taylor, Cheryl Black, Chris Burns, Antigoni Ekonomou, and Peter Braude. 2005. Generation of a human embryonic cell line encoding the cystic fibrosis mutation ^F508, using preimplantation genetic diagnosis. *Reproductive Biomedicine Online* 10 (3): 390–397.

Plomer, Aurora. 2011. EU ban on stem cell patents is a threat both to science and the rule of law. *Guardian* December 12.

President's Commission for the Study of Bioethical Issues. 2010. *New Directions: The Ethics of Synthetic Biology and Emerging Technologies.* http://bioethics.gov/cms/studies.

President's Commission for the Study of Bioethical Issues. 2011a. *Research across Borders: Proceedings of the International Research Panel of the Presidential Commission for the Study of Bioethical Issues.* http://bioethics.gov/cms/studies.

President's Commission for the Study of Bioethical Issues. 2011b. *"Ethically Impossible" STD Research in Guatemala from 1946 to 1948*. http://bioethics.gov/cms/studies.

President's Commission for the Study of Bioethical Issues. 2011c. *Moral Science: Protecting Participants in Human Subjects Research*. http://bioethics.gov/cms/studies.

President's Council on Bioethics. 2003. *Beyond Therapy: Biotechnology and the Pursuit of Happiness*. US Independent Agencies and Commissions.

President's Council on Bioethics. 2005. *Alternative Sources of Human Pluripotent Stem Cells*. US Independent Agencies and Commissions.

President's Council on Bioethics. 2008. *Human Dignity and Bioethics: Essays Commissioned by the President's Council on Bioethics*. US Independent Agencies and Commissions.

Puar, Jasbir. 2007. *Terrorist Assemblages: Homonationalism in Queer Times*. Duke University Press.

Rabinow, Paul. 1992. Artificiality and enlightenment: From sociobiology to biosociality. In *Incorporations*, ed. Jonathan Crary and Sanford Kwinter. Zone.

Rabinow, Paul. 2008. *Marking Time: On the Anthropology of the Contemporary*. Princeton University Press.

Rabinow, Paul, and Gaymon Bennett. 2009. Synthetic biology: Ethical ramifications 2009. *Systems and Synthetic Biology* 3 (1–4): 99–108.

Rader, Karen. 2004. *Making Mice: Standardizing Animals for American Biomedical Research, 1900–1955*. Princeton University Press.

Rao, Radhika. 2005. Coercion, commercialization, and commodification: The ethics of compensation for egg donors in stem cell research. *Berkeley Technology Law Journal* 21 (3): 1055–1066.

Rapp, Rayna. 2000. *Testing Women, Testing the Fetus: The Social Impact of Amniocentesis in America*. Routledge.

Rawls, John. 2001. *Justice as Fairness: A Restatement*. Harvard University Press.

Reardon, Jennifer. 2011. The "persons" and "genomics" of personal genomics. *Personalized Medicine* 8 (1): 95–107.

Reardon, Jennifer, and Kim TallBear. 2012. "Your DNA is our history": Genomics, anthropology, and the construction of whiteness as property. *Current Anthropology* 53 (suppl. 5): S233–S245.

Rei, Wenmay, and Terence Hua Tai. 2010. Introduction: Biotechnology in East Asian societies: Controversies and governance. *EASTS* 4 (1): 1–6.

Revazova, E., N. Turovets, O. Kochetkova, L. Kindarova, L. Kuzmichev, J. Janus, and M. Pryzhkova. 2007. Patient-specific stem cell lines derived from human parthenogenetic blastocysts. *Cloning and Stem Cells* 9 (3): 432–449.

Reverby, Susan, ed. 2000. *Tuskegee's Truths: Rethinking the Tuskegee Syphilis Study.* University of North Carolina Press.

Reynolds, Jesse. 2010. Bending the rules in California. *Biopolitical Times,* June 16. http://www.biopoliticaltimes.org/article.php?id=5261.

Richards, Evelleen. 1997. Redrawing the boundaries: Darwinian science and Victorian women intellectuals. In *Victorian Science in Context,* ed. Bernard Lightman. University of Chicago Press.

Riles, Annelise, and Marie-Andrée Jacob. 2007. The new bureaucracies of virtue: Introduction. *Political and Legal Anthropology Review* 30 (2): 181–191.

Robert, Jason Scott. 2006. The science and ethics of making part-human animals in stem cell biology. *FASEB Journal* 20: 838–845.

Roberts, Dorothy. 1999. *Killing the Black Body.* Vintage.

Roberts, Dorothy. 2003. *Shattered Bonds: The Color of Child Welfare.* Basic Books.

Roberts, Dorothy. 2011. *Fatal Invention: How Science, Politics, and Big Business Re-Create Race in the Twenty-First Century.* New Press.

Rockstanding. 2009. Save rats, kill people? GE looking for ways to further research. http://rockstanding.com/2009/07/09/save-rats-kill-people-ge-looking-for-ways -to-further-research.

Rodriguez, Juana. 2003. *Queer Latinidad: Identity Practices, Discursive Spaces.* NYU Press.

Rose, Hilary. 2007–2008. From Nuremburg to informed consent in the 21st century. *Review of Bioethics* 1 (1): 20–37.

Rose, Nikolas. 2006. *The Politics of Life Itself: Biomedicine, Power, and Subjectivity in the Twenty-First Century*. Princeton University Press.

Rosenberg, Charles E. 1999. Meanings, policies, and medicine: On the bioethical enterprise and history. *Daedalus* 128: 27–46.

Roth, Duane. 2012. *CIRM's IP Policies: Ensuring Innovation and Fair Return*. California Institute for Regenerative Medicine.

Royal Society. 2012. *Science as an Open Enterprise*.

Russell, W., and R. L. Burch. 1959. *The Principles of Humane Experimental Technique*. http://altweb.jhsph.edu/publications/humane_exp/het-toc.htm.

Saad, Lydia. 2009. More Americans "Pro-Life" Than "Pro-Choice" for First Time. Gallup Politics (http://www.gallup.com/poll/118399/more-americans-pro-life-than-pro-choice-first-time.aspx).

Saha, Krishanu, and Benjamin Hurlbut. 2011. Treat donors as partners in biobank research. *Nature* 478: 312.

Salari, Keyan, Philip Pizzo, and Charles Prober. 2011. To genotype or not to genotype? Addressing the debate through the development of a genomics and personalized medicine curriculum. *Academic Medicine* 86 (8): 925–927.

Salter, Brian. 2010. Governing Innovation in the biomedicine knowledge economy: Stem cell science in the USA. *Science & Public Policy* 37 (2): 87–100.

Salter, Brian, and Olivia Harvey. 2008. Stem cell innovation in the USA: The benefits of the minimal state. *Regenerative Medicine* 3 (4): 597–610.

Saxton, Marsha. 2000. Why members of the disability community oppose prenatal diagnosis and selective abortion. In *Prenatal Testing and Disability Rights*, ed. Erik Parens and Adrienne Asch. Georgetown University Press.

Scheper-Hughes, Nancy. 2005. The last commodity: Post-human ethics and the global traffic in "fresh" organs. In *Global Assemblages*, ed. Aihwa Ong and Stephen Collier. Blackwell.

Scheper-Hughes, Nancy. 2006. The tyranny of the gift: Sacrificial violence in living donor transplants. [Ethics Corner.] *American Journal of Transplantation* 7: 1–5.

Scheper-Hughes, Nancy, and Loic Wacquant, eds. 2002. *Commodifying Bodies*. Sage.

Schicktanz, Silke, Mark Schweda, and Brian Wynne. 2011. The ethics of "public understanding of ethics"—why and how bioethics expertise should include public and patients' voices. *Medicine, Health Care, and Philosophy* 15: 129–139.

Schlesinger, Arthur. [1949] 1988. *The Vital Center: The Politics of Freedom.* Da Capo.

Schrag, Brian. 2006. Research with groups: Group rights, group consent, and collaborative research. *Science and Engineering Ethics* 12: 511–521.

Scott, Christopher. 2006. *Stem Cell Now.* Penguin.

Scott, Christopher. 2008. What stem cell therapy can learn from gene therapy. *Nature Reports Stem Cells.* Published online September 4.

Scott, David. 2003. *Wall Street Words: An A to Z Guide to Investment Terms for Today's Investor.* Houghton Mifflin.

Scott, James. 1992. *Domination and the Arts of Resistance: Hidden Transcripts.* Yale University Press.

Seifert, Christie. 1997. Comment: Fetal tissue research: State regulation of the donation of aborted fetuses without the consent of the "mother." *John Marshall Law Review* 31 (1): 277–298.

Shakespeare, Tom, and Nicholas Watson. 2002. The social model of disability: An outdated ideology? *Research in Social Science and Disability* 2: 9–28.

Shanks, Pete. 2012. Stem cell fraud is the real issue in Texas. *Biopolitical Times*, March 2. http://www.biopoliticaltimes.org/article.php?id=6099.

Shapin, Steven. 2008. *The Scientific Life: A Moral History of a Late Modern Vocation.* University of Chicago Press.

Shea, Dana. 2007. *Oversight of Dual-Use Biological Research: The National Science Advisory Board for Biosecurity.* Congressional Research Service.

Shiva, Vandana, and Ingunn Moser, eds. 1995. *Biopolitics: A Feminist and Ecological Reader on Biotechnology.* Zed.

Sigurdsson, Skúli. 2002. Bioethics Lite™: Two aspects of the health sector database deCODE controversy. In *Schöne-Gesunde-Neue Welt? Das Humangenetische Wissen und Seine Anwendung aus Philosophischer, Soziologischer und Historischer Perspektive,* ed. T. Hornschuh, K. Meyer, G. Rüve, and M. Voß. Bielefeld University.

Silliman, Jael, Marlene Fried, Loretta Ross, and Elena Gutierrez. 2004. *Undivided Rights: Women of Color Organizing for Reproductive Justice*. South End.

Silver, Lee. 2004. The God Effect: America's religious conservatives aren't the only ones who object to science on spiritual grounds—so do Europe's Greens. The big winner is Asia. *Newsweek International*, April.

Simoncelli, Tania, and Barry Steinhardt. 2006. California's Proposition 69: A dangerous precedent for criminal DNA databases. *Journal of Law, Medicine & Ethics* 34 (2): 199–213.

Simpson, Bob. 2013. Managing potential in assisted reproductive technologies: Reflections on gifts, kinship, and the process of vernacularization. *Current Anthropology* 54, Supplement 7.

Singer, Peter. 2001. *Animal Liberation*. Harper Perennial.

Singer, Peter. 2011. *Practical Ethics*, third edition. Cambridge University Press.

Sinnott-Armstrong, Walter. 2011. Consequentialism. In *The Stanford Encyclopedia of Philosophy* (winter 2011 edition), ed. Edward Zalta. http://plato.stanford.edu/archives/win2011/entries/consequentialism/.

Sipp, Douglas. 2009. Stem cell research in Asia: A critical view. *Journal of Cellular Biochemistry* 107 (5): 853–856.

Skloot, Rebecca. 2010. *The Immortal Life of Henrietta Lacks*. Crown.

Slaughter, Sheila, and Gary Rhoades. 2009. *Academic Capitalism and the New Economy: Markets, State, and Higher Education*. Johns Hopkins University Press.

Smaglik, Paul. 2000. Tissue donors use their influence in deal over gene patent terms. *Nature* 407: 891.

Smaglik, Paul. 2003. Filling biopolis. *Nature* 425: 746–747.

Smith, Kimberly. 2007. *African American Environmental Thought: Foundations*. University of Kansas Press.

Smith, Linda Tuhiwai. 1999. *Decolonizing Methodologies: Research and Indigenous Peoples*. Zed.

Smith Hughes, Sally. 2011. *Genentech: The Beginnings of Biotech*. University of Chicago Press.

Snow, C. P. 1959. *The Two Cultures and the Scientific Revolution*. Cambridge University Press.

Somers, Terri, 2006. As moral debate sidetracks stem cell research in the U.S., countries in Asia, Europe are moving to stake claims in the promising industry. *San Diego Union-Tribune*, December 17.

Spar, Deborah. 2006. *The Baby Business: How Money, Science, and Politics Drive the Commerce of Conception*. Harvard Business Review Press.

Squier, Susan. 2005. *Liminal Lives: Imagining the Human at the Frontiers of Biomedicine*. Duke University Press.

Stadtfeld, M., M. Nagaya, J. Utikal, G. Weir, and K. Hochedlinger. 2008. Induced pluripotent stem cells generated without viral integration. *Science* 322 (5903): 945–949.

Steinbrook, Robert. 2006. Egg donation and human embryonic stem-cell research. *New England Journal of Medicine* 354 (4): 324–326.

Steinmann, Michael, Peter Sykora, and Urban Wiesing, eds. 2009. *Altruism Reconsidered: Exploring New Approaches in Human Tissue*. Ashgate.

Stephan, Paula. 2012. *How Economics Shapes Science*. Harvard University Press.

Stern, Alexandra. 2005. *Eugenic Nation: Faults and Frontiers of Better Breeding in Modern America*. University of California Press.

Stevens, Tina, and Diane Beeson. 2006. Opinion: A Closer Look at Stem Cell Research. *Oakland Tribune*, January 18.

Stoler, Ann Laura. 2009. *Along the Archival Grain: Epistemic Anxieties and Colonial Common Sense*. Princeton University Press.

Stone, Christopher. 1974. *Should Trees Have Standing? Toward Legal Rights for Natural Objects*. William Kaufmann.

Strathern, Marilyn. 2005. *Partial Connections*, updated edition. Altamira.

Strathern, Marilyn. 2011. An end and a beginning for the gift? *Journal de la Société des Océanistes* 130–131: 119–128.

Streiffer, Robert. 2008. Informed consent and federal funding for stem cell research. *Hastings Center Report* 38 (3): 40–47.

Sunder Rajan, Kaushik. 2006. *Biocapital: The Constitution of Postgenomic Life*. Duke University Press.

Sunder Rajan, Kaushik, ed. 2012. *Lively Capital: Biotechnologies, Ethics, and Governance in Global Markets*. Duke University Press.

Sunstein, Cass. 2004. Can animals sue? In *Animal Rights: Current Debates and New Directions*, ed. Cass Sunstein and Martha Nussbaum. Oxford University Press.

Svendsen, Mette, and Lene Koch. 2008. Unpacking the "spare embryo": Facilitating stem cell research in a moral landscape. *Social Studies of Science* 38 (1): 93–110.

Swearengen, James, ed. 2005. *Biodefense: Research Methodology and Animal Models*. CRC Press.

Takahashi, K., and S. Yamanaka. 2006. Induction of pluripotent stem cells from mouse embryonic and adult fibroblast cultures by defined factors. *Cell* 126: 663–676.

Takahashi, Kazutoshi, Koji Tanabe, Mari Ohnuki, Megumi Narita, Tomoko Ichisaka, Kiichiro Tomoda, and Shinya Yamanaka. 2007. Induction of pluripotent stem cells from adult human fibroblasts by defined factors. *Cell* 131 (Issue 5): 861–872.

Talan, Jamie. 2010. Federal policy allows funding for new human embryonic stem cell lines: But not all investigators are happy with the changes. *Neurology Today* 10 (3): 6.

TallBear, Kimberly. 2012. *Native American DNA: Origins, Race, and Governance*. University of Minnesota Press.

Taussig, Karen-Sue. 2009. *Ordinary Genomes: Science, Citizenship, and Genetic Identities*. Duke University Press.

Tayag, Joseph. 2006a. Toward fair cures: Diversity policies in stem cell research. The Greenlining Institute Issue Brief #1, October. http://greenlining.org/resources/pdfs/TowardFairCuresDiversityPoliciesinStemCellResearch.pdf

Tayag, Joseph. 2006b. Economic development and stem cell research. The Greenlining Institute Issue Brief #2, October. http://greenlining.org/resources/pdfs/EconomicDevelopmentandStemCellResearch.pdf

Teman, Elly. 2010. *Birthing a Mother: The Surrogate Body and the Pregnant Self*. University of California Press.

Terry, Jennifer. 2000. "Unnatural acts" in nature: The scientific fascination with queer animals. *GLQ: A Journal of Lesbian and Gay Studies* 6 (2): 151–193.

Thompson, Charis. 2002a. When elephants stand for competing philosophies of Nature. In *Complexities: Social Studies of Knowledge Practices*, ed. John Law and Annemarie Mol. Duke University Press.

Thompson, Charis. 2002b. Ranchers, scientists, and grass-roots development in the United States and Kenya. *Environmental Values* 11 (3): 303–326.

Thompson, Charis. 2004. Co-producing CITES and the African elephant. In *States of Knowledge: The Co-Production of Science and Social Order*, ed. Shiela Jasanoff. Routledge.

Thompson, Charis. 2005. *Making Parents: The Ontological Choreography of Reproductive Technologies*. MIT Press.

Thompson, Charis. 2006. Race science. In *Theory, Culture & Society* 23 (2–3), *Special Issue on Problematizing Global Knowledge*, ed. Mike Featherstone, Couze Venn, Ryan Bishop, and John Phillips.

Thompson, Charis, 2007a. Can opposition to research spur innovation? *Nature Reports: Stem Cells* (December 13), *Featured Editor Commentary*.

Thompson, Charis. 2007b. Why we should, in fact, pay for egg donation. *Regenerative Medicine* 2 (2): 203–209.

Thompson, Charis. 2007c. God is in the details: Reproductive technologies and religion. Special Issue: *Culture, Medicine and Psychiatry* 30 (4): 557–561.

Thompson, Charis. 2008. Stem cells, women, and the new gender and science. In *Gender Innovations in Science and Engineering*, ed. Londa Schiebinger. Stanford University Press.

Thompson, Charis. 2009. Skin tone and the persistence of biological race in egg donation for assisted reproduction. In *Shades of Difference*, ed. Evelyn Nakano Glenn. Stanford University Press.

Thompson, Charis. 2010a. Asian regeneration? Nationalism and internationalism in stem cell research in South Korea and Singapore. In *Asian Biotech*, ed. Aihwa Ong and Nancy Chen. Duke University Press.

Thompson, Charis. 2010b. Informed consent for the age of pluripotency and embryo triage: From alienation, anonymity, and altruism to connection, contact, and care. In *The "Healthy" Embryo*, ed. Jeff Nisker et al. Cambridge University Press.

Thompson, Charis. 2011. Medical migrations afterword: Science as a vacation? *Body & Society* 17 (2&3): 205–213.

Thompson, Charis. 2012. Biomaterialien und Bioinformationen nehman, Berkeley USA 2010. In *Bioökonomie: die Lebenswissenschaften und die Bewirtschaftung der Körper*, ed. Susanne Lettow. Transcript.

Thompson Cussins, Charis. 1999. Confessions of a bioterrorist: Subject position and reproductive technologies. In *Playing Dolly: Technocultural Formations, Fantasies, and Fictions of Assisted Reproduction*, ed. E. Ann Kaplan and Susan Squier. Rutgers University Press.

Thomson, James, et al. 1998. Embryonic stem cell lines derived from human blastocysts. *Science* 282 (5391): 1145–1147.

Tierney, John. 2007. Are scientists playing God? It depends on your religion. *New York Times*, November 20.

Till, J. E., and E. A. McCulloch. 1961. A direct measurement of the radiation sensitivity of normal mouse bone marrow cells. *Radiation Research* 14: 213–222.

Tolley, David C. 2008. Regulatory priorities governing stem cell research in California: Relaxing revenue sharing and safeguarding access plans. *Berkeley Technology Law Journal* 23: 219.

Trager, Rebecca. 2009. US funding boost—A threat to EU science? Royal Society of Chemistry, April 6. http://www.rsc.org/chemistryworld/News/2009/April/06040903.asp.

Trinh, Minh-ha. 2005. *The Digital Film Event*. Routledge.

Trivers, Robert. 1971. The evolution of reciprocal altruism. *Quarterly Review of Biology* 46: 35–57.

Trommelmans, Leen, J. Selling, and K. Dierickx. 2007. A Critical Assessment of the Directive on Tissue Engineering of the European Union. *Tissue Engineering* 13 (4):667–672.

Turner, Leigh. 2007. From Durham to Delhi: "Medical tourism" and the global economy. In *Comparative Program on Health and Society Lupina Foundation Working*

Papers Series 2006–7, ed. Jillian Cohen-Kohler and M. Bianca Seaton. Munk Centre for International Studies, University of Toronto.

Tutton, Richard, and Oonagh Corrigan, eds. 2004. *Genetic Databases: Socio-Ethical Issues in the Collection and Use of DNA*. Routledge.

Twyman, Richard. 2002. What are model organisms? *The Human Genome*. London: Welcome Trust. http://genome.wellcome.ac.uk/doc_WTD020803.html.

United Nations Non-Governmental Liaison Service. 2003. *Intergovernmental Negotiations and Decision-Making at the United Nations*.

University of California. 2010. Diversity annual accountability sub-report. http://www.universityofcalifornia.edu/diversity/documents/diversity-accountability-report-and-appendix-0910.pdf

Visvanathan, Shiv. 1997. *A Carnival for Science: Essays on Science, Technology, and Development*. Oxford University Press.

Vora, Kalindi. 2009. Indian transnational surrogacy and the commodification of vital energy. *Subjectivity* 28: 266–278.

Wadman, Meredith. 2010. Breast cancer gene patents judged invalid. *NATNews* 30 (March). http://www.nature.com/news/2010/100330/full/news.2010.160.html.

Wailoo, Keith. 1997. *Drawing Blood: Technology and Disease Identity in Twentieth-Century America*. Johns Hopkins University Press.

Wailoo, Keith, Alondra Nelson, and Catherine Lee. 2012. *Genetics and the Unsettled Past: The Collision of DNA, Race, and History*. Rutgers University Press.

Wald, Priscilla. 2008. *Contagious: Cultures, Carriers, and the Outbreak Narrative*. Duke University Press.

Waldby, Catherine. 2008. Oocyte markets: Women's reproductive work in embryonic stem cell research. *New Genetics & Society* 27 (1): 19–31.

Waldby, Catherine. 2009. Singapore biopolis: Bare life in the city-state. *East Asian Science, Technology, and Society: An International Journal* 3: 367–383.

Waldby, Catherine, and Melinda Cooper. 2010. From reproductive work to regenerative labour: The female body and the stem cell industries. *Feminist Theory* 11 (1): 3–22.

Waldby, Catherine, and Robert Mitchell. 2006. *Tissue Economies: Blood, Organs and Cell Lines in Late Capitalism.* Duke University Press.

Washington, Harriet. 2008. *Medical Apartheid: The Dark History of Medical Experimentation on Black Americans from Colonial Times to the Present.* Harlem Moon.

Washington, Harriet. 2011. *Deadly Monopolies: The Shocking Corporate Takeover of Life Itself—And the Consequences for Your Health and Our Medical Future.* Doubleday.

Watson, J. D., and F. H. C. Crick. 1953. A structure for deoxyribose nucleic acid. *Nature* 171 (3): 737–738.

Watt, Nicholas. 2006. US faces science brain drain after Europe backs stem cell funding. *Guardian,* July 25, p. 17.

Weindling, Paul. 2001. The origins of informed consent: The International Scientific Commission on medical war crimes, and the Nuremberg Code. *Bulletin of the History of Medicine* 75 (1): 37–71.

Welsome, Eileen. 2000. *The Plutonium Files: America's Secret Medical Experiments in the Cold War.* Delta.

White House. 2012. *National Bioeconomy Blueprint.*

Williams, Patricia. 2012. How our genetic maps are being sold to the highest bidder. *Nation,* June 25. http://www.thenation.com/article/168261/how-our-genetic-maps-are-being-sold-highest-bidder#.

Wilmut, Ian, Keith Campbell, and Colin Tudge. 2000. *The Second Creation: Dolly and the Age of Biological Control.* Farrar, Straus & Giroux.

Wilmut, I., et al. 2002. Somatic cell nuclear transfer. *Nature* 419: 583–586.

Wilson, Ara. 2011. Foreign bodies and national scales: Medical tourism in Thailand. *Body & Society* 17 (2–3): 121–137.

Wilson, Duncan. 2011. Creating the "ethics industry": Mary Warnock, in vitro fertilization, and the history of bioethics in Britain. *Biosocieties* 6: 121–141.

Winickoff, David. 2006. Governing stem cell research in California and the USA: Towards a social infrastructure. *Trends in Biotechnology* 24 (9): 390–394.

Winickoff, David, Krishanu Saha, and Gregory Graff. 2009. Opening stem cell research and development: A policy proposal for the management of data, intellectual property, and ethics. *Yale Journal of Health Policy, Law, and Ethics* 9 (1): 52–127.

Wolf, Susan, et al. 2012. Managing incidental findings and research results in genomic research involving biobanks and archived data sets. *Genetics in Medicine* 14 (4): 361–384.

Wolinsky, Howard. 2009. Home is where the bench is. *EMBO Reports* 10: 1196–1198.

Woltjen, Knut, Iacovos Michael, Paria Mohsen, Ridham Desai, Maria Mileikovsky, Riikka Hamalainen, Rebecca Cowling, et al. 2009. *piggyBac* transposition reprograms fibroblasts to induced pluripotent stem cells. *Nature* 458: 766–770.

Wong, Kai Wen, and Tim Bunnell. 2006. "New economy" discourse and spaces in Singapore: a case study of one-north. *Environment and Planning* 38:69–83.

Woo Cummings, Meredith, ed. 1999. *The Developmental State.* Cornell University Press.

Xu, Chunhui, M. S. Inokuma, J. Denham, K. Golds, et al. 2001. Feeder-free growth of undifferentiated human embryonic stem cells. *Nature Biotechnology* 19: 971–974.

Yoon, Jeong-Ro, Sung Kyum Cho, and Kyu Won Jung. 2010. The challenges of governing biotechnology in Korea. *East Asian Science, Technology and Society: An International Journal* 4 (2): 335–348.

You, May-Su, and Vladimir Korzh. 2005. Zebrafish in the Tropical One-North. *Zebrafish* 1 (4): 327–334.

Young, Emma. 2002. California challenges US stem cell rules. *New Scientist*, September 23.

Young, Iris. 2000. *Inclusion Democracy.* Oxford University Press.

Yu, Junying, et al. 2007. Induced pluripotent stem cell lines derived from human somatic cells. *Science* 318 (5858): 1917–1920.

Yudof, Mark. 2008. UC President Mark Yudof statement on Santa Cruz firebombings. http://www.universityofcalifornia.edu/news/article/18342.

Yusa, Kosuke, S. Tamir Rashid, David Lomas, Allan Bradley, Ludovic Vallier, et al. 2011. Targeted gene correction of α1-antitrypsin deficiency in induced pluripotent stem cells. *Nature* 478(7369):391-394.

Zarkov, Dubravka. 2007. *The Body of War: Media, Ethnicity, and Gender in the Break-up of Yugoslavia*. Duke University Press.

Zelizer, Viviana. 2010. *Economic Lives: How Culture Shapes the Economy*. Princeton University Press.

Zhao, Xiao-yang, W. Li, Z. Lv, L. Liu, et al. 2009. iPS cells produce viable mice through tetraploid complementation. *Nature* 461 (3): 86–90.

INDEX